What Brain Research Teaches about Rigor, Relevance, and Relationships
and What It Teaches about Keeping Your Own Brain Healthy

Willard R. Daggett and Paul D. Nussbaum

Excepting those portions intended for classroom or training use, no part of this publication may be reproduced in whole or in part, or stored in a retrieval system, or transmitted in any form or by any means, electronic, mechanical, photocopying, recording, or otherwise, without written permission of the publisher. For information regarding permission, write to International Center for Leadership in Education. The International Center for Leadership in Education grants the purchaser of this publication permission to reproduce those pages intended for use in classrooms or training. Notice of copyright must appear on all copies of copyrighted materials.

Copyright © 2008 by International Center
for Leadership in Education

All rights reserved.
Published by International Center for
Leadership in Education.
Printed in the U.S.A.

ISBN-0-9656553-7-7

International Center for Leadership in Education
1587 Route 146 • Rexford, New York 12148-1137
(518) 399-2776 • fax (518) 399-7607
info@LeaderEd.com • www.LeaderEd.com

#B-08-BRR

Contents

Preface v

About the Authors vii

I. The Emergence of Brain Health 1

II. Integration of Rigor and Relevance with the Human Brain 43

III. Brain Research — The Basis of Successful Practices 67

Appendix 1 Your Personal Brain Health Survey 95

Appendix 2 Introduction to Food and Your Brain 105

Appendix 3 Resources 127

Preface

As a lifelong educator and as a parent and now a grandparent, I have always been fascinated by student learning. How does the brain work that enables learning to occur? Why do some children and adults seem to learn quicker and more easily than others? Why do various teaching techniques work better with some students than others?

These and similar questions led me to study the growing body of research on the brain. This research base has exploded recently due to the increase capacity and sophistication of imaging technology.

As I looked at this complex and expanding research, I found myself getting lost in the technical details. Then I came across the work of Dr. Paul D. Nussbaum.

Dr. Nussbaum has a unique capacity to make the complex simple. He has the ability to bring what is written in the medical research journals down to a practical level that can help me as a parent/grandparent and educator. Therefore, I reached out to Paul and asked him if he would be willing to work with the International Center to put together a publication so we could share his insights with a wide variety of audiences.

As Paul and I began to sort through the implications of brain research on learning, it became apparent that in many ways I had a myopic view of the research's importance. I was studying the brain research simply to see how I could help teachers become more effective in the classroom and parents become more effective in helping their children learn.

Paul showed me that while teaching children more effectively was an important goal, it was far too limited. Brain research has revealed how to create a healthy brain for a lifetime. It is this broader priority that is so critical to preparing young people for success in life.

The knowledge we provide students about brain health and the lifetime habits we help them develop in their youth will determine more than just how well they do in school. Far more important, that knowledge can influence the choices they make, the practices in which they engage, and the commitments they make in pursuit of a healthy, prosperous, and engaging life.

Paul has captured the essence of these goals. His work has influenced my life greatly, and I believe it will influence yours as well.

Willard R. Daggett, Ed.D.
President
International Center for Leadership in Education

About the Authors

Willard R. Daggett, Ed.D.

Dr. Willard R. Daggett, President of the International Center for Leadership in Education, is recognized worldwide for his proven ability to move education systems towards more rigorous and relevant skills and knowledge for *all* students. He has assisted a number of states and hundreds of school districts with their school improvement initiatives, many in response to *No Child Left Behind* and its demanding adequate yearly progress (AYP) provisions. Dr. Daggett has also collaborated with education ministries in several countries and with the Council of Chief State School Officers, the Bill & Melinda Gates Foundation, the National Governors Association, and many other national organizations.

Before founding the International Center for Leadership in Education in 1991, Dr. Daggett was a teacher and administrator at the secondary and postsecondary levels and a director with the New York State Education Department. Dr. Daggett is the author of six books about learning and education, 12 textbooks and numerous research studies, reports, and journal articles.

Dr. Daggett has spoken to hundreds of thousands of educators and education stakeholders in all 50 states. His enlightening, entertaining, and motivating messages have helped his listeners to look at education differently by challenging their assumptions about the purposes, benefits, and effectiveness of American schools. Dr. Daggett inspires his audiences both to embrace what is best about our education system and to make the changes necessary to meet the needs of *all* students in the 21st century.

www.LeaderEd.com • info@LeaderEd.com • (518) 399-2776

Paul D. Nussbaum, Ph.D.

Dr. Paul D. Nussbaum is a licensed clinical neuropsychologist in Pennsylvania. Having earned his doctorate in clinical psychology from the University of Arizona in 1991, Dr. Nussbaum completed his internship and post-doctoral fellowship at Western Psychiatric Institute and Clinic, University of Pittsburgh School of Medicine. He is an adjunct Associate Professor in Neurological Surgery at the University of Pittsburgh School of Medicine and an international consultant on aging and brain health.

Dr. Nussbaum has 20 years experience in the care of older persons suffering dementia and related disorders. From the outpatient setting to the long-term care setting, Dr. Nussbaum has worked in all sectors of the continuum of care. An expert in neuroanatomy and human behavior, Dr. Nussbaum has published many peer reviewed articles, books, and chapters within the scientific community. He is a national and international lecturer on brain health, healthy aging, dementia, and related disorders.

Dr. Nussbaum educates the general public on the basics of the human brain and how to keep the brain healthy over the entire lifespan. He provides informative and entertaining keynote presentations across the United States and is often interviewed by regional and national press/media. Dr. Nussbaum also provides consultation to schools and corporate America on brain health.

www.paulnussbaum.com • drnuss@zoominternet.net
(412) 471-1195

Section I

The Emergence of Brain Health

Paul D. Nussbaum, Ph.D.

You have no greater asset than your life story. It must be shared with your next great generation, the little ones in your life.

During the past decade I have traveled the nation speaking about the mystery and miracle of the human brain. My talks are delivered to groups and organizations both small and large. From basements of small churches to the National Press Club and beyond, the enthusiasm for learning about the human brain has been remarkable. It is personally pleasing to see the increase in popularity of the human brain and the emergence of a serious directive to develop health policy and programs for this essential part of our being. We are truly engaged in a cultural shift toward brain health that has already witnessed new products, new programs, and new companies geared to the human brain.

Our society is undergoing a not-so-quiet revolution regarding the human brain and brain health. The concept of brain health is now discussed in major sectors of society including personal development, health care, business, media, and even religion. It is common to see information on brain health in major news outlets, popular magazines, peer-reviewed medical journals, business periodicals, initial public offerings, and even in a new television series. This is good news because it indicates a cultural shift in which the United States is prepared and willing to begin the process of integrating brain health into our language and more importantly into our daily health regimen.

I have had the unique opportunity to travel the nation the past decade discussing the basics of the human brain and teaching audiences from all backgrounds what the research suggests we can do to improve our brain health. Americans want to learn about the brain and how to maintain brain health. Some of this enthusiasm may be steeped in fear, or a real concern over losing mental ability and memory as we age. The enthusiasm, however, is also based on a desire to remain healthy. Perhaps most gratifying is that the message of brain health resonates with persons from all backgrounds. People want information that will benefit their health, and our society is more invested than ever before in a brain healthy lifestyle.

The human brain is the single greatest, most magnificent system ever designed in the history of the universe! Most Americans, however, do not know the basics of the brain and therefore cannot take care of the brain. A recent survey of Americans on brain health sponsored by MetLife Foundation and the American Society on Aging (www.asaging.org) found only 3% identified brain health as a leading health topic. More promising, the same survey found a majority (88%) believe they can keep their brains fit and nearly

The Emergence of Brain Health

90% believe regular brain checkups are important. Americans also demonstrate an understanding of what activities are considered good for brain health. In order for our nation to be enlightened about brain health we need to better educate our citizens about the human brain and I begin this education early in life.

In the United States, there has been much focus on educating the public about the importance of cardiac or heart health. Heart health is seen as a priority because many people have family members with cardiac illness and is a primary cause of death. Some research indicates that, while cardiac illness remains a leading cause of death, there has been some progress in slowing the rate of cardiac-caused premature death. Persistent education campaigns teach the public about behaviors that promote cardiac health. Aerobic centers, exercise clubs, and television programs on exercise are all popular. Grocery stores and restaurants have sections dedicated to heart-healthy foods. Small red heart icons are now used on some packaging to identify heart-healthy food.

Our culture has also adopted a general fondness of the heart as a favorite organ in our body. Like the ancient Egyptians, we have a belief that our being revolves around the heart. Our language contains many statements that give the heart meaning it really does not deserve. For example, statements such as "I love you with all my heart," "the Steelers played their hearts out," and "you broke my heart" suggest our heart has the capacity for emotion or feelings. Indeed, the human heart is a pump that perfuses blood throughout the body. It has no emotion like love, thought, or motor skill. All of the emotions and thoughts that we relate to the heart are really owned by the brain. In this regard, our brains have not been treated fairly, and I believe it is time we begin to give the brain the attention it deserves.

At this period in our nation's history, we have an unprecedented opportunity to be part of a societal shift toward brain health that will likely lead to an unleashing of human potential and maybe a reduction in brain disease. Some of our advances may occur in the development of traditional medical interventions such as the development of a vaccine or a new medication therapy. Gene therapy and stem cell research with use of our own stem cells to combat disease may offer a new frontier of treatment or prevention options. The other major advancement is a new national priority to focus on the human brain and declare our desire to address the loss of family members to brain disease. Similar to our approach to cardiac health, the United States can become an enlightened society on brain health by implementing a national lifelong educational program on the basics of the human brain. Preventative programs can adopt and pay for a brain health lifestyle for all of us and the development of brain health centers can offer research-based activities for the consumer. Grocery stores and restaurants can begin the process of identifying for the consumer foods that have brain health-promoting effects. Businesses that cater to the vitality of the human brain and our cognitive/emotional abilities will continue to emerge.

A nation that prioritizes brain health understands that a lifelong dedication to brain health and a proactive lifestyle are needed. Individuals must educate themselves about their own brains and begin to make the behavioral changes necessary to develop as healthy a brain as possible. Corporate America, businesses, media and television, and health care systems can promote brain health

The Emergence of Brain Health

in their own specialized ways, and our daily language will reflect a society that embraces the importance of caring for the brain. This is my third book to champion brain health as a national priority and to continue my effort to educate America about the importance of basic brain education. I believe you will care for your brain if you learn the basics of your brain.

This book will educate you about this miraculous part of your being. You will have fun reading this book because you will take interest in yourself. The book provides you with a proactive lifestyle for brain health. The lifestyle encompasses five important components and the research-based activities that can be organized within each component. As you read the book, think about yourself and your current thoughts about your brain. Review your current lifestyle and make critical decisions regarding necessary changes to promote the health of your brain.

Once you have read this book, you will know more about the human brain than most Americans. More importantly, you will know what behaviors are important to promote your own brain health. This knowledge will help you to take the steps to change your current behavior and adopt a proactive lifestyle for brain health. It is not easy and there is no "quick fix." Rather, it is a difficult process filled with self-inspection and challenge. I have found personal satisfaction in my own behavioral change as I work to integrate brain health into my life. I also have fun with others as we work together to adopt brain health changes. Remember, you are working on yourself. It is time for you to consider your brain and how to keep it healthy!

Cool Stuff About Your Brain

Your brain weighs 2-4 pounds and is made up of gray matter and white matter. The gray matter tends to be contained in an area of your brain called the *cortex*, a word that literally means "bark of a tree." Your cortex (see Figure 1) is a convoluted mass of cells with folds and flaps that sits snug within your skull.

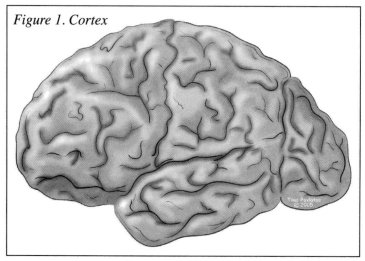

Figure 1. Cortex

Tina Pavlatos © 2005

The cortex developed from the back to the front, meaning the front part of the cortex is the youngest region of the brain. The cortex is primarily responsible for your most complex thinking abilities including memory, language, planning, concept formation, problem solving, spatial representation, auditory and visual processing, mood, and personality. Processing in the cortex tends to be conscious and intentional.

The cortex is generally organized into four primary regions or lobes: frontal lobe, temporal lobe, parietal lobe, and occipital lobe (see Figure 2). Each of these four lobes has specific behaviors and functions primary to its region. For example, the frontal lobe is also known as the *executive system,* since it helps to execute and organize behavior and plans, conceptualizes and maintains cognitive flexibility and mood stability. Your personality is

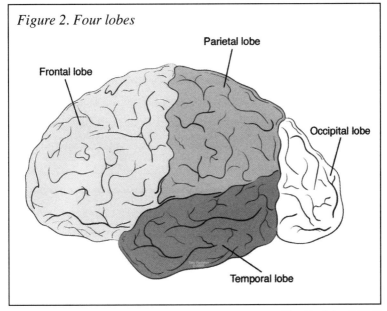

Tina Pavlatos © 2005

thought to reside in the frontal region of the brain. The temporal lobe is the site of your auditory brain, memory and new learning, language, and perhaps religiosity. The parietal lobes help you with orientation to space, memory, reading and writing, mathematics, and appreciation of left versus right. Finally, the occipital lobes help you to see, discriminate what you see, and to perceive.

Sitting just under the cortex and on top of the ascending brain stem are a number of smaller and generally more primitive structures (relative to the cortex) known as the *subcortex*. The subcortex primarily processes rote skills and procedures. Some, if not most, of the processing conducted in the subcortex is subconscious. Functions such as driving, dressing, typing, etc., involve multiple rote procedures that are conducted at a subconscious level. The subcortex and cortex are distinct regions of the brain, but they do not sit in isolation of one another. In fact, there are numerous connections between these two important brain regions. The brain operates as a symphony with numerous and distinct regions harmonizing perfectly as one unit.

Another way to learn about your brain is to consider that it has two sides, one called the left hemisphere and other called the right hemisphere (see Figure 3).

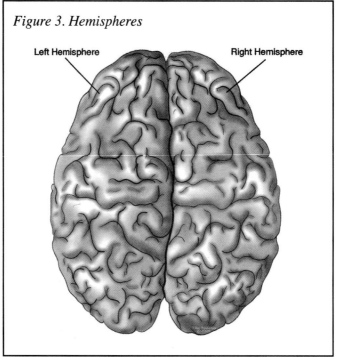

Figure 3. Hemispheres

The Emergence of Brain Health

Interestingly, your behaviors and functions are related primarily to one of these two hemispheres. For example, most of us and nearly all right handers have language distributed primarily in the left hemisphere. We refer to the hemisphere with language as the *dominant hemisphere* as a sign of our respect to the importance of language. Left-handed persons with a parent who is left handed (relatively rare) have a higher probability than the right handers of having language distributed primarily in the right hemisphere. They would be *right hemisphere dominant.*

Your dominant hemisphere (left for most of us) also processes details, is task oriented, is logical and analytical, and sequences information. Most of western civilization is built around the left hemisphere, for example, as our classrooms are set up in rows and columns of chairs and our cities tend to have tall building in rows and columns. We tend to focus more on the detail and less on the gestalt. Your *non-dominant hemisphere* helps you process non language information such as size, shapes, sounds, and space. Your ability to navigate in space, locate your car in a parking garage, and return home from a walk is an example of non-dominant function. Likewise, your ability to appreciate distinct sounds such as a baby's cry or a fire alarm tends to be a process of the non-dominant hemisphere.

Your two hemispheres are connected by a bridge of cells called the *corpus collosum*. Information crosses from one side of your brain to the other over the corpus collosum and this is a critical part of your brain's ability to remain so functional despite its many complex operations on a daily basis. Interestingly, the female brain is thought to have a larger corpus collosum and underscores the notion that female brains process information differently from male brains. Females tend to utilize both sides of their brains more to process whereas men rely primarily on one side, the dominant hemisphere. It is probably not coincidental that audiences across the nation always respond with the same answer to my question, "What is a common behavior that men and women struggle with on a daily basis?" The answer is communication!

The operation and function of your brain is ultimately conducted by millions of brain cells called *neurons*. A neuron (see Figure 4) contains a cell body sometimes referred to as a *soma,* a long arm called an *axon* extending out from the cell body and branch-like figures called *dendrites* that extend out into the brain environment seeking new information to relay back to the cell body.

Information from the cell body travels down the axon into the surrounding brain while information from the environment is gathered by the dendrites and brought back to the cell body. This ongoing exchange of information by the brain is why we refer to it as the *central information processing system.*

We are taught that our brains contain millions of brain cells and that each neuron can communicate with another 10,000 neurons. Interestingly, one neuron never touches another neuron, but two cells may communicate via chemicals. This chemical connection is called a *synapse.* The more synaptic connections you develop over your lifespan the healthier your brain may be because it is building up *brain reserve*. Brain reserve, as you will learn about later, may have the ability to delay the onset of neurodegenerative diseases such as Alzheimer's disease (AD).

The Emergence of Brain Health

The miracle is that your brain is dynamic and continues to be shaped and to develop — it has *plasticity*. As such, there is no finite capacity or limitation. In this way your brain is distinct and much superior to a computer because computers will always have built-in limitations and finite capacity. Your dynamic brain is shaped by environmental input across your lifespan beginning in the womb. There really is no critical period of brain development unless one considers life itself to be the measure. As you will learn in the next section of this book, the type of environmental input to your brain can make a difference in the health of your brain. You do have some control and this is great news!

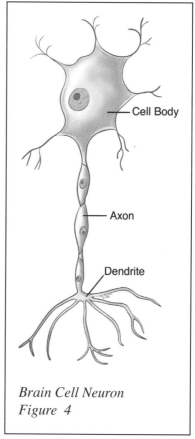

Brain Cell Neuron
Figure 4

Tina Pavlatos © 2005

When I give lectures to the public I always want my audience to personalize the message. This story is, after all, about you and your brain. It really does not get any more personal. Learning about oneself can be fun and challenging. There is one part of your brain that I emphasize because this structure, *the hippocampus,* is so critical to you and your life story. The hippocampus (see

Figure 5) sits in the middle of each temporal lobe that lies under the temples on each side of your head.

Your hippocampi (plural for hippocampus), as you have one in each hemisphere, take new information in and maintains

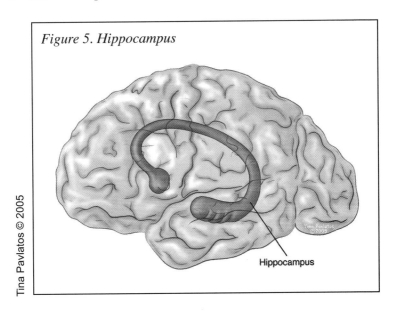

Figure 5. Hippocampus

the information in a type of working buffer. If you believe the information is important and you need to store the information for an extended period of time, the hippocampus will transition the information to a specific area of the cortex. This process is not random, but very sophisticated as the process of storage seems to be stimulus-based. That is, if you are learning information that is visual your hippocampus will help store that information permanently in the visual cortex of the brain. The same process is thought to occur for the other four types of sensory input.

The Emergence of Brain Health

The hippocampi represent your vital learning and encoding structures thereby helping you to build your life story and maintain your personal memories. AD is a leading cause of dementia and the disease destroys the hippocampi of the brain very early. As a result, those afflicted with this terrible brain disease cannot learn new information. As you will learn in the next section of this book, the hippocampi are critical structures to the story of brain health. Recent research suggests that the hippocampi have tremendous ability including new brain cell development referred to as *neurogenesis*.

You now know more than most Americans about the human brain. You have now personalized this critical part of your being and you should feel empowered and excited to learn more. Let us review some important information about the basics of your brain:

- Your brain weighs 2-4 pounds.
- Your brain is comprised of at least 60% fat. It is the fattiest system in your body.
- Every heartbeat provides 25% of the blood and oxygen to your brain.
- Your brain has a cortex and a subcortex.
- Your brain has a left hemisphere (primarily language) and a right hemisphere (non-language functions).
- Your hippocampi encodes new information and initiates learning and memory.
- You have millions of brain cells (neurons) that can be shaped and increased in number with exposure to complex and novel environments.

What Brain Research Teaches about RRR

- Neurons communicate chemically with each other, and this is referred to as a synapse.
- The more synaptic connections, the greater your brain reserve.
- Brain reserve is thought to delay the onset of diseases such as AD.

How Brain Health Works

Now that you have learned about your brain I hope you understand the unique asset and miracle that sits within your own skull. Your two-to-four-pound brain is the true portable and wireless device in the universe! You have learned about the structure or hardware of your brain. Now it is important to learn about the power of your brain and how it can be shaped and nurtured across your entire lifespan. We refer to the dynamic, constantly reorganizing, and malleable nature of your brain as *brain plasticity*. Your brain is not a rigid or static system with a limited capacity or finite critical period for development. The power of brain plasticity permits you to implement a lifelong and proactive program to grow and promote your own brain health.

Environment and the Rat Brain

To better understand why brain plasticity is important to you we can discuss the basic findings of animal brain research. In the 1950s, research was conducted to investigate whether environment had any effect on the structure and function of the animal brain. Researchers designed a study with rodents raised in two distinct environments: enriched environment versus unenriched environment. Rodents were raised in one of these two environments and then their brains were analyzed and compared at autopsy. Results yielded significant differences in the brains of these rodents. Specifically, rodents raised in an enriched environment had a larger cortex, more cellular connections (a synapse that leads to brain reserve), and new brain cells (called neurogenesis) in the hippocampus (structure critical to new learning and memory).

I became interested in this work and wanted to know how researchers defined an enriched environment. My review of this work (see *Brain Health and Wellness*, 2003) suggested three factors were critical to the enriched environment. They are (1) *socialization*: animals had to have other animals of its kind in the environment; (2) *physical activity*: animals had a running wheel to exercise on; and (3) *mental stimulation*: there were toys in the environment that animals could play and interact with. Animals raised in unenriched environments were raised in isolation, had no running wheel, and had no toys. While this research offered highly significant findings regarding the effect of the environment on brain structure in the animal, the critical issue of whether the same findings could be established for humans remained unknown.

Environment and the Human Brain

In 1998, a landmark study found the human brain had the ability to develop new brain cells. This study was a threshold moment for our species, as it confronted our long tradition of believing the human brain to be a rigid system with no ability to generate new brain cells. We had always believed the brain is born with all of its brain cells, that the human brain loses brain cells on a daily basis, and that our brains do not generate or replace the lost cells with new ones. The study also indicated that the new brain cells were generated in the human hippocampus — the same area neurogenesis was found in the animal brain. Today, research is ongoing to determine if neurogenesis occurs in other regions of the human brain or if it is specific to the hippocampus.

New brain cell development is one outcome of a brain with plasticity. Recall that plasticity refers to a brain that is dynamic, constantly reorganizing, and malleable. The human brain, therefore, is now thought to possess the same type of neural plasticity as the rodent brain. Interestingly, the animal studies were conducted on rodents across their lifespan with an equivalent human age of 70 or 80. A human brain that generates new brain cells mandates a curiosity of how this wonderful adaptive ability occurs. We can return to the animal studies to derive some answers to this question. Recall that the enriched environment led to new brain cell development in the hippocampus of the animal. The three critical factors important to the enriched environment, as discussed earlier, are socialization, physical activity, and mental stimulation. It makes good sense, therefore, to ask if the human brain can be similarly affected by environment and if the enriched environment promotes positive brain changes in the human.

As you will read in the next section there is good reason to believe that the human brain benefits from a novel and complex environment. It is also important to know that the first potential enriched environment is the womb and that the type of environment you expose your brain to will have consequences at every age across your lifespan. The miracle of brain plasticity does not end at a particular age. Indeed, the human brain probably does not know its chronological age and will demand and benefit from enriched environments at every age. The major point of this section is that you are strongly encouraged and empowered to expose your brain to the novel and complex every day regardless of your age!

The Importance of Brain Reserve

Brain reserve is a well-known concept that refers to a build up of brain cell connections (*synaptic density*) that serves to assist the brain in the battle against neurodegenerative diseases. To better understand brain reserve, consider the following simplistic analogy that I use in my lectures on the human brain. Imagine flying in an airplane nearly 1,000 feet above the ground. As you peer out your window down to the ground you will see two distinct scenes. The first scene is a jungle in which there are so many trees you cannot see the ground. The second scene is an island with one palm tree blowing slowly in the wind. You want your brain to be like the jungle in which you have a tremendous number of synaptic connections (*synaptic density*). This is a direct measure of brain reserve. You do not want your brain to look like the island with one palm tree. The reason is simple. AD and other types of dementia act like a weed whacker cutting through the plants around your house. If your brain looks like a jungle, filled with synaptic connections, it will take dementia a long time to show its ugly face. However, if your brain looks like the island with one palm tree, the clinical signs of dementia will manifest quickly because there is no reserve to fight it off.

Indeed, some research has shown that even though brains are diagnosed with AD at autopsy due to the presence of neuropathological changes, a significant number of these persons never demonstrated the clinical aspects of the disease in life. This is explained using the brain reserve concept. That is, persons who never manifested AD in life, even though they had the neuropathologic characteristics in their brain at autopsy, had built up brain reserve to fight off or delay the onset of the disease.

The power of brain reserve is further supported by findings that relate higher education and occupation levels to a lower risk of AD. For those with high education or occupation levels who do manifest AD, their presentation of the disease occurs later than those without similar backgrounds, and once the disease manifests, the person dies soon after. The reason is that when the disease presents clinically it is already advanced into the final stage because the person's brain reserve had been fighting it.

Education and occupation are two examples of environments that can be enriched. You expose your brain to education and occupation environments frequently across your lifespan. Each of these two environments provides the opportunity for you to engage in a novel and complex setting that promotes the development of brain reserve. To the extent that these environments or other settings become rote and passive, brain reserve will not be as developed and the overall health benefit for your brain is not as great. It is your personal challenge to expose your brain to a novel and complex environment on a daily basis. Studies suggest the earlier in life you begin your exposure to the enriched environment, the greater the health benefit to your brain, even into late life. This finding is supported by research that demonstrates a relationship between poverty in childhood and increased risk of AD later in life, higher IQ in childhood and young adulthood and reduced risk of AD later in life, language development in young adulthood and reduced risk of neuropathologic changes in the brain at autopsy, and a passive lifestyles in the 40s to increased risk of AD later in life.

These findings on humans support the idea that diseases of the brain that manifest late in life may actually begin early in life. Further, these findings suggest we can become involved very early in life with a proactive lifestyle that promotes brain health and that helps to reduce the risk of AD and related dementias later in life. It is important to prioritize a proactive lifestyle for brain health regardless of your age, to embrace the power of brain plasticity and development of brain reserve, and to have fun in the process of caring for your brain!

What have you learned in this section? Here is a quick review:

- The animal brain has the ability to generate new brain cells, particularly in the hippocampus.
- The animal brain demonstrates positive effects from being exposed to an enriched environment.
- The enriched environment for the animal includes socialization, physical activity, and mental stimulation.
- In 1998, we learned the human brain also has the ability to generate new brain cells. Like animals, humans generate new brain cells in the hippocampus.
- Your brain, therefore, has plasticity and the ability to be shaped by environment at any age.
- By exposing your brain to the enriched environment (the novel and complex), you build up your brain reserve.
- Brain reserve is thought to fight off or delay the onset of brain disease.

What Brain Research Teaches about RRR

- It is important for you to engage in a proactive lifestyle that promotes brain health to maintain your life story!

Your Brain Health Lifestyle Program

Your brain is a highly dynamic and constantly reorganizing system capable of being shaped across your entire lifespan. Similar to animals, your brain can generate new brain cells and respond to environmental input. Your goal is to expose your brain to enriched environments — to the novel and complex — and to grow your brain reserve! Stimuli that is considered rote and passive to your brain is most likely not as health promoting. We learned from animal brain research that an enriched environment has three critical factors: socialization, physical activity, and mental stimulation. These same critical factors are important to the human brain. As part of my work on human brain health I have proposed a lifestyle that includes five critical factors (three from the animal research and two additional factors):

1. Socialization
2. Physical Activity
3. Mental Stimulation
4. Spirituality
5. Nutrition

Each of these factors is necessary to your brain health program and together they form an integrated whole for you. The five factors need to be understood as one program and not separate entities. Remember, your goal is to adopt a proactive lifestyle for brain health that increases your brain reserve through exposure to the complex and novel. As with any lifestyle program, the journey can be difficult and perhaps humbling. There is no easy or quick answer. Your brain health program is a lifelong journey towards wholeness and will require constant personal review and change. While this program is not easy, the goal of a healthier and more challenged brain is worthwhile. I have found that spending time inspecting and reviewing my own behavior, and making an effort to adopt healthy change in my lifestyle, can be fun.

Getting Started and Taking Inventory

The first step for your brain health program is to understand the five critical parts of the program: socialization, physical activity, mental stimulation, spirituality, nutrition. Also, you must think of your brain as a highly dynamic system that will react to the types of input you feed it. From this perspective, you can appreciate how much control you have regarding the potential health of your brain.

It is helpful to review your current lifestyle to better understand the positive and negative aspects for your brain health. You will find the **Brain Health Survey** in the appendix. Derive your *Baseline Score* for the Inventory prior to starting your Lifestyle Program. It is important to be honest and to understand this is simply a guide to give you an idea of your *Baseline Brain Health Lifestyle*. Do not be alarmed if you give your current lifestyle a low grade. You have not been educated by society about the importance of your brain and you have not been informed about brain health. That is about to change!

What Brain Research Teaches about RRR

This section explores the five parts of Your Brain Health Lifestyle Program. You have already taken the time to learn about your brain and how capable it is at every age. You know your brain reacts favorably to enriched environments that promote the novel and complex. This includes growth of new brain cells and an increase in your brain reserve. You have also taken the time to review your current lifestyle and you applied an honest grade for your brain health prior to starting your program (Baseline Brain Health Lifestyle).

Empowered by information on your brain and an honest assessment of your current brain health lifestyle, you are ready to learn what activities are believed to promote brain health. Research-based activities or behaviors are organized and presented for each of the five critical parts of your brain health lifestyle program. As you read the following section, it is important that you think about why such activities promote brain health and whether you have these activities in your current lifestyle. Most importantly, what will you need to do or change to include these activities in your daily lifestyle to maximize your brain health?

Socialization and Brain Health

Human beings need to be with other human beings. We really do not have a choice regarding this fact as it has been in our DNA since the beginning of time. Socialization was one of the critical factors of the enriched environment for animals that helped to foster healthy brain development. Research teaches us that humans who isolate or segregate have a higher risk of dementia than those who remain integrated in society. Dementia is a clinical term that refers to loss of general intelligence, memory deficit, loss of other thinking abilities, personality change, and functional decline. There are more than 70 causes of dementia, and AD is the leading cause of dementia in the United States.

Socialization and brain health might be explained by the opportunity for communication, critical thought, creativity, emotional expression including intimacy, chemical connection, touch, and recreation that arises when two or more humans interact. Personal meaning and identity might also be a result of interpersonal activity or the creation of an entity or mission "larger than oneself."

> **Brain Health Lifestyle Tip:** Stay involved in your community at every age, do not retire, and have a personally meaningful reason for getting up each day!

I had the unique opportunity to provide a brief presentation on *Brain Health in America* at the National Press Club in September 2006 (see asaging.org). I voiced my opposition to our national policy of retirement since it contradicts nearly everything about brain health. A nation enlightened on brain health encourages active involvement across the lifespan and does not reinforce or encourage removal of oneself from society to a passive and potentially isolated environment. As I travel the nation teaching audiences about their brains, I always underscore the importance of remaining involved in meaningful ways. I discourage retirement as some identify their worth and very being by their occupation! What happens psychologically when they no longer have a job, particularly in a society like ours that still has archaic policies of mandatory retirement?

Since we live in a society that still has mandatory retirement for some occupations, I believe the development of hobbies beginning in early to middle life (20s to 60s) is important. A hobby represents a brain that has been challenged. Multiple hobbies reflect a robust brain with neural networks that have been nurtured. Development of hobbies is a highly important behavior and a challenge for the baby boomers (those born between 1946 and 1964). Hobby development creates an enriched environment and provides a vehicle for the brain to experience the novel and complex. What hobbies do you have and do you have interests that you have been resisting or putting off for some time? Take one such interest and get started today. You are on your way to building brain reserve!

A Practical Exercise to Promote Socialization

Every community has a variety of clubs, organizations, and formal groups that seek membership. These may be part of a local church, school, or community. While most of these memberships require volunteer time, they provide the value of socialization and contribution to an ongoing enterprise. What will it take for you to explore the opportunities in your community where you can provide input and value? It starts by first understanding what skills you have. Most Americans do not know their real mission on the planet. We tend to be too busy to think about such questions or to explore such issues. It is an interesting question that requires some deep thought and time. If you have discovered your true mission, the opportunity exists to align what it is you are called to do with what you actually do. Happiness and productivity are typically the outcome of such alignment.

The Emergence of Brain Health

You have a wonderful list of talents that probably has not been tapped. Take a few moments and prepare a list of talents you think you possess. It does not matter if you have expressed them yet. These talents are most likely not related to your occupation or job description and they probably represent those things that you would like to pursue "if I only had the time." Once you have made your list of talents or skills, begin to relate them to the list of organizations or clubs in your community. Do you notice any potential alignments where your talents can increase the value of the particular organization or club? You might even have the entrepreneurial spirit to begin your own club, group, or business using your talents to lead the way! The point of this exercise is to realize that socialization is important to brain health, that identifying opportunities for socialization in your community and combining that with your own innate talents can foster an enriched environment for your brain health.

> **Brain Health Tip:** Develop hobbies, identify your own innate talents, and align them with ongoing groups or organizations in your community.

Physical Activity and Brain Health

Animals that ran on a wheel generated new brain cells in studies conducted in the late 1950s. This research underscores the importance of physical activity to animal brain health. The same relationship between physical activity and brain health appears to be true for humans. It is important to understand why physical activity relates to brain health. Every time your heart beats, 25% of the blood and nutrients from that one heartbeat goes directly to your brain. We have known for some time that physical exercise is critical to cardiac health. Research is now beginning to underscore a similar value for physical exercise to brain health! A 2006 study by Colcombe and colleagues found that as little as three hours a week of brisk walking (aerobic exercise) increases blood flow to the brain and may trigger neurochemical changes that increase the production of new brain cells. The regions of the brain most affected by the aerobic exercise included the frontal lobes (important for complex thinking, reasoning, and attention) and the corpus collosum (the bundle of white matter that bridges the two sides of the brain). This study is important for several reasons:

1. The results further support brain plasticity and new brain cell development in humans.
2. The study was conducted on persons aged 60-79 indicating brain health can occur in later life. This is consistent with animal brain research showing positive effects at advanced ages. Remember, I do not believe in a critical period of brain development unless it is defined as life!
3. This may be the first study to demonstrate healthy structural changes in the human brain with physical activity — a finding we know exists for animals.
4. We know a relationship exists between physical changes in the brain (positive and negative) and functional or cognitive ability.

I have no doubt that research will further support the relationship between physical activity and brain health. This relationship likely exists at all ages and with healthy and diseased brains. Do not underestimate the power of blood flow and oxygen to the brain. Other research suggests walking on a daily basis or at least several times a week can reduce the risk of dementia. This finding again supports the relationship between physical activity and reduction in the risk of brain disease. Reduction in the risk of dementia is what I call "brain health." Interestingly, there appears to be a dosing effect; the more you walk during the week, the more positive effect for the brain. I have learned that most Americans know they should be walking daily and that they can even specify the need to walk about 10,000 steps daily. This tells me that our nation has done a good job educating the consumer. Unfortunately, education does not necessarily translate into action or behavior. I read recently that only about 35% of our nation is involved in a formal and consistent exercise program! As one interested in behavior, I always want to provide a personal touch to behavior and behavior change. I can advise you to walk on a daily basis and to take 10,000 steps daily. However, what are the chances that you will actually do it?

My recommendation to my audiences interested in brain health is to purchase a pedometer. You can purchase a pedometer at any local shopping mall or sports store. You will derive tremendous value for your $15.00 purchase as the pedometer will keep track of your daily steps and will also remind you to walk. I always recommend you buy one for a loved one in your family; it makes a great birthday present. Have some fun with it!

> **Brain Health Tip:** Get physically active with at least three hours of aerobic exercise a week and walk for distance three to five times per week. It is recommended that we all walk around 10,000 steps daily! Purchase a pedometer.

What Brain Research Teaches about RRR

You now have learned that aerobic exercise and walking on a daily basis provide physical and functional benefits to your brain. You know why physical activity helps to increase brain health (25% of blood output from each heartbeat) and you have been given a practical tip on purchasing a pedometer to help change your behavior. You will be pleased to learn that there are other physical activities you can enjoy that relate to reduced risk of dementia. The interesting thing about these activities is that you will need to use both sides of your body, a brain-boosting exercise. I am often amused by the fact that most of us not only have a dominant side, but also have almost completely neglected our non-dominant side. It is important to understand that each side of your body is controlled by the opposite side of the brain. As such, most of us have essentially ignored one half of the brain! My message to audiences across the nation interested in brain health is to start on the path to becoming ambidextrous; an ambidextrous brain is a healthier brain.

Other brain-health promoting physical activities include dance, particularly the tango as it has been shown to reduce the risk of dementia. I am not sure we have the ability to specify how much dance or how often we should dance yet, but this behavior appears to be healthy for the brain. Gardening and knitting are two activities that also relate to reduced risk of dementia. Once again, notice that dance, gardening, and knitting demand use of both sides of the body. In thinking about how knitting and gardening might lead to brain health, it is useful to consider what the brain is asked to do with these activities. For example, with gardening your brain will be asked to plan into the future, engage in visuospatial function

The Emergence of Brain Health

(where to plant the corn relative to the carrots), and visuomotor skill. This says nothing about the stress reduction effect gardening might offer. We Americans need to learn that a health effect can be derived by things other than pills, liquids, and shots!

> **Brain Health Tip:** Consider taking dance lessons, starting a garden, and learning how to knit.

There are some general rules that appear to be useful regarding physical activity and brain health:

1. Cardiovascular health is important to brain health. The more you can increase the strength of your heart and the output of blood from the heart through physical activities, the healthier your brain will probably be.
2. Focus your behavioral change on those physical activities research has found to be important to brain health.
3. Begin to develop an ambidextrous brain by using your non-dominant body half more often. Consider writing with your non-dominant hand several minutes every day. You will be amazed as your practice leads to increased comfort and legibility (brain reserve is being built).

Mental Stimulation and Brain Health

Because your brain is the single greatest information processing system in the universe it is not surprising that many people focus on the mental stimulation factor in brain health. There are numerous computer-based products that aim to provide memory and other cognitive training exercises. You can challenge yourself with these mental exercises on a daily basis with the hope of improving your different cognitive or thinking abilities. You can use the software on your own computer or gain access to an Internet site, where you can complete your mental workout. A 2006 study by Willis and colleagues is the first to document long-term positive effects of cognitive training on everyday function in older adults. We will likely continue to see new businesses emerging around the desire to improve the mental aspects of the brain. This is another example of a cultural shift towards brain health!

At the same time new businesses are developing, research is providing information on what mentally stimulating lifestyle activities help to promote brain health or to reduce the risk of dementia by building brain reserve. Language appears to be critical regarding brain development and sophistication of the language system in young adulthood might actually be predictive of brain health in late life. Dr. Snowdon, who leads the Nun Study, has found that the number of ideas expressed in a diary written by 21-year-old females predicted the percentage of tangles in the brain (marker of AD) nearly 60 years later. Snowdon proposed that language sophistication in early life might mark a well-developed brain that is resistant to neurodegenerative changes later in life. In contrast, a language system that is not well developed in early life may mark a vulnerable brain at risk for neurodegenerative changes in later life.

This work is supported by other studies that report a relationship between IQ in early life and the risk of dementia such as AD in later life. Young boys who fought in World War II took an IQ test prior to their enlistment. Their scores on the IQ test at the age of 18 were found to correlate with the presence of AD in later

The Emergence of Brain Health

life. The higher the IQ at age 18, the lower the risk of AD and vice versa. Other research in Scotland has found a relationship between mental state scores at age 7 and cognitive integrity in the 70s. These findings underscore the point made earlier about the need to be proactive and treat brain health as a lifespan issue.

There is some interesting work on sign language in infants prior to their neurological ability to speak. Infants can learn about 20 signs prior to being able to speak orally. When the infants exposed to sign language are followed as they grow, they show greater articulation abilities and their IQ is higher by the 2nd grade relative to controls. As we learned earlier, higher IQ early in life relates to reduced risk of dementia later in life. Once again, interventions early in life that enhance IQ and develop the language system appear to be examples of proactive brain health. The good news is we know IQ can be increased with good nutrition, a loving environment, breast feeding, and sign language. How many of these behaviors are in your health care plan? Every baby wellness program or Head Start program should include each of these brain health behaviors!

Brain Health Tip: Develop your language system, learn a new language, read and write daily, and expose your brain (particularly your baby's) to sign language.

Research has taught us that playing boardgames helps to reduce the risk of dementia. You probably never thought of your family game of Scrabble or Monopoly as a brain health workout. Other games such as Poker, Bridge, Sudoku, and crossword puzzles probably promote brain health as long as they are "novel and complex." Once any activity becomes rote and passive, the positive brain health effects are reduced.

What Brain Research Teaches about RRR

Reading and writing on a daily basis are good for your brain. These activities help promote new learning and, thereby, involve the hippocampus. The more you stimulate and massage your hippocampi, the better. Try to read new material or new topics and write with the intent of expressing ideas. Remember to write with your non-dominant hand a few minutes a day. Reading and speaking to the baby developing in the womb may have neurological benefits to the baby. You are encouraged to get involved in shaping your baby's brain!

Lifelong learning programs are now part of the social norm in many universities across the nation that sponsor such programs. I have had the pleasure of learning about the value and fun of Elderhostel and Osher Lifelong Learning programs. Hundreds of thousands of older adults are enrolling in university classroom work as part of their "retirement." What used to be the beach or the golf course is now a book and a classroom! I have spoken for many years about how learning is a health-promoting entity that is no different than a medication. Learning involves structural, chemical, and functional changes in the brain that can be health promoting. Indeed, research indicates education is a major factor contributing to longevity and health. The actual event of learning something new involves the laying down of a new neural network that was not there before. With continued learning, the brain develops a rich network of neural associations, or brain reserve. It is this brain reserve that helps to delay the onset of neurodegenerative disorders such as AD. For this reason, I have argued lifelong learning programs should be part of every health care payer system, including Medicare!

The Emergence of Brain Health

Lifelong learning means learning throughout the lifespan and does not infer a starting point in late life. I believe every elementary school in this nation should be mandated to teach the basics of the brain to American children. In order to be proactive, we need to begin very early in life. Our children will more likely care for their brains if they understand this wonderful part of their being. The earlier the education occurs, the earlier a proactive lifestyle for brain health can be started. As we learned earlier, the types of environments we expose our brains to early in life relate to the health of our brains later in life. In this regard, I am pleased to know that teachers and neuroscientists are working together to merge the lesson plan and the brain scan!

Brain Health Tip: Enroll in a lifelong learning program in your community or at your local University. Encourage your local school board to integrate curriculum on the basics of the human brain within the elementary school — be proactive!

Classical music has been found to have a relationship to learning in children. It is not uncommon to observe classical music being played in some classrooms during study period or even during a test. Study time at home can be enhanced with background classical music. Once again, some work indicates classical music played for the baby developing in the womb has neurological benefits. You are also encouraged to learn to play a musical instrument. It is true that learning to play a musical instrument is harder as you age, but your brain can learn an instrument at any age. Do not be afraid to develop that dormant part of your brain!

> **Brain Health Tip:** Start to play those wonderful board games again, learn a musical instrument, and tune in to the classical radio station.

Travel has been shown to reduce the risk of dementia, or to increase brain health. Consider how this behavior might promote brain health. You already know that the best environment for your brain is the complex and novel. When you travel away from home you are leaving a familiar surrounding and exposing your brain to a novel and complex environment. As a result, you will use your cortex to navigate and you will probably find it both exciting and frustrating. Interestingly, as you stay in that new environment you will become more familiar and comfortable. The novel and complex has become rote and passive. Everyday you travel to and from work and home and essentially do not use your cortex. Your subcortex has the mental maps of your home and neighborhood and the processing tends to be rote or subconscious. New environments are more brain health promoting. Travel also allows you to meet new people who contribute to your enriched environment.

> **Brain Health Tip:** Try to take a trip or two this year to a new surrounding and enjoy the brain health benefits!

Spirituality and Brain Health

I have noticed that Americans are a bit timid when talking about spirituality. Often an audience member will approach me after my talk and whisper a thank-you for speaking about spirituality. I am not sure how our nation got to this point, but there will be no fear in this text to discuss an important health-promoting behavior. Spirituality has many meanings and it may mean something different to you than to me. This section refers to spirituality as one means of turning inward to a peaceful existence and to remove oneself from the hurried society that is America. In this regard, I employ prayer, meditation, and relaxation procedures as three tangible examples of spirituality.

Animal research has found that rodents raised in an environment that is too stimulating demonstrate slowed brain development. In addition, animals exposed to environments that are highly stressful and where they have little control demonstrate structural damage in their hippocampus and evince memory problems. The point of this animal research is that humans should pause and examine how fast we are moving on a daily basis and decide if we need to slow down. Early research on the human brain exposed to life-threatening stressors indicates there is similar damage to the human hippocampus as is known to exist in animals. Also, humans with chronic anxiety have memory problems that again support the negative effect of stress and uncontrolled anxiety on brain function.

Research and surveys have reported the following positive effects of prayer on health:

1. Prayer on a daily basis relates to an enhanced immune system — the system that helps you defend against colds, flu, and other illnesses.
2. people who attend a formalized place of worship live longer and report happier and healthier lives than those who do not.

Surveys suggest only 30% of the nation attend a formalized place of worship weekly.
3. Prayer as part of the daily routine while in the hospital relates to an earlier discharge. I would think the health care payer system would find this interesting!
4. According to a past *Parade Magazine* survey, 95% of the physicians in the United States believe prayer is important to the well-being of their patients. I wonder if a prescription for prayer makes sense.

> **Brain Health Tip:** Consider attending a formalized place of worship on a weekly basis and incorporate prayer into your daily health routine.

Meditation and relaxation procedures are also good techniques to help you slow down and to turn inward for balance and symmetry. Your brain can adapt to a chaotic world, but it will function more efficiently over a longer period of time if you provide moments of inward reflection and rest. Meditation offers one technique to achieve such inner peace. Interestingly, the brains of monks in deep meditation show changes in glucose metabolism. The brains of the monks no longer distinguish the outside world from their internal world. The saying "I became one with my world" takes on real meaning. Most Americans do not know how to meditate and part of your brain health program can include a lesson or two on meditation so you can engage in this behavior on a daily basis.

Similar to meditation, relaxation procedures that include deep breathing and progressive muscle relaxation exercises are not well known or used by Americans. Our nation tends to favor the pill over lifestyle behaviors, the quick fix over patience. We tend to use band-aid approaches and really do not fix the underlying problems or struggles of our lives. It is important to identify what part of your body is vulnerable to stress. You may experience stress in your neck, lower back, head, or stomach. By identifying what part of your body is your stress target site you can then engage in progressive muscle relaxation procedures to alleviate the stress from your target site. For example, squeeze your right hand into a fist as hard as you can. Hold the fist and pay attention to how uncomfortable the tension is in your fist. Now, slowly release your fingers extending them and notice how the tension leaves your fingers. The more relaxed feeling achieved by letting the fist go and extending your fingers is an example of how you can focus on any muscle group in your body to release stress. You can actually tighten and release muscles all over your body two to three times a day. At the end of the exercise you will notice that energy in the form of stress has left you and you will feel better!

Similarly, you may not know how to breathe correctly or to use breathing techniques to rid your body of stress. You are encouraged to engage in proper relaxation breathing exercises two to three times daily. Taking a deep inhalation through your nose using your stomach muscles and holding the breath for several seconds will result in your feeling some tension in your stomach and chest. Now, slowly release the air from your stomach and chest to your mouth in a rhythmic way. Doing this exercise several times throughout the day can help you to slow down and gain a sense of calm as you rid yourself of toxic stress.

> **Brain Health Tip:** Enroll in a meditation class and begin to incorporate meditation into your brain health program. Employ progressive muscle relaxation and deep breathing exercises two to three times daily. Turning inward will help your brain escape temporarily the stress filled and unenriched environments of life.

One final brain health-promoting behavior you will enjoy is sleep! A large number of Americans have sleep disorders and it is common for Americans to respond "I am tired" when they are asked how they are doing. Sleep is actually a very active time for the brain. There are four stages of sleep and the brain requires each one, particularly rapid eye movement (REM), the 25% of sleep when you dream. If you do not sleep, your brains will sleep for you, which can be fatal, such as when one falls asleep while driving.

Some studies indicate that your brain actually consolidates information into a well-organized and formed memory during sleep. You can appreciate why sleep is so important if it is involved in your information processing and ability to recall potentially important details. Those who do not get enough sleep have a higher risk of thinking problems and mood disorder such as depression. Work-related accidents and motor vehicle accidents are examples of negative outcomes from poor sleep. Are you getting enough sleep? If not, why not and what are you doing about it? There are many reasons for sleep disorders including anxiety, bills, pain, interpersonal or family conflict, and medication side effects. Each of these causes can have a different solution or treatment to regain a normal sleep pattern. Remember that sleep is a learned behavior and can be altered in many ways.

A sleep pattern can be relearned with consistent behavior regarding time to go to bed, using the bed for sleep and not reading or watching TV, and not lying in bed when you are not sleeping. You can probably give yourself 20 to 25 minutes to try and get to sleep in your bed. If you cannot fall asleep after that time period, get out of the bed and leave the bedroom. You may even identify an "anxiety chair" in your house where you can sit and worry. When you are done worrying go back to your bed and try to sleep. The same 20 to 25 minute rule holds and you may need to get up several times. The point of this exercise is to condition your body and brain to sleep in the bed, not worry in the bed!

> **Brain Health Tip:** Try to get enough sleep to feel rested in the morning. Speak to your health care provider if you have sleep problems to understand why and to develop a remedy.

Nutrition and Brain Health

Brain health and nutrition has become a very popular and intense area of study. Entire books are devoted to food and its effect on the brain. A new journal, *Nutritional Neurosciences*, recognizes this emerging specialization. This book cannot cover the entire content area of brain health and nutrition, but it is included and represents a critical part of my Brain Health Lifestyle Program.

What Brain Research Teaches about RRR

General tips regarding brain health and nutrition include the following:

- Increase your intake of omega-3 fatty acids because your brain contains at least 60% fat.
- Increase your intake of antioxidants to help combat free radical development.
- Increase your intake of colored fruits and vegetables including green leafy vegetables.
- Eat with utensils and you will likely eat healthier foods and reduce your caloric intake.
- Reduce your intake of red meat.
- Eliminate trans-fatty acids from your diet and try to reduce your intake of processed foods.
- Eat at least one meal a day with family or friends. This is a rich brain health behavior because you communicate, slow down, eat with utensils, eat healthy foods, and eat less.

Integration of Rigor and Relevance with the Human Brain

Daily Brain Health Activities

Physical Activity	Mental Stimulation	Spirituality	Socialization	Nutrition
• Walk daily • Aerobic exercise • Dance • Garden • Knit • Use non-dominant side • Jog • Purchase a pedometer • Biking, swimming • Light weights	• Board games • Read and write • Sign language • Computer-based products • Travel • Novel and complex • Develop language skills • Learn a musical instrument • Listen to classical music	• Daily prayer • Meditate • Relaxation techniques • Increase sleep • Learn to say "no" • Slow down • Yoga and Pilates • Attend formal place of worship • Identify body targets of stress	• Hobbies • Join a new group • Dine with others • Do not retire • Grow network of friends • Recreate and laugh • Be forgiving • Connect with family • Personal mission in life	• Salmon • Herring and mackerel • Walnuts • Colored fruits • Colored vegetables • Increase antioxidants • Eat 80% of portion served • Use utensils • Green leafy vegetables

These are suggestions for activities to incorporate into your brain health lifestyle. As with any healthy lifestyle, disease can still occur.

Section II

THE INTEGRATION OF
RIGOR AND RELEVANCE
WITH THE HUMAN BRAIN

Paul D. Nussbaum, Ph.D.

Brain Health and Education: A Historical Context

My journey toward brain health in general and the more specific focus of this section, brain health and education, began nearly 10 years ago. As a licensed neuropsychologist I am well trained in neuroanatomy, brain behavior relations, and the clinical aspects of conditions that affect the brain such as stroke and Alzheimer's disease (AD). This training and clinical experience provided me a foundation of understanding from a disease perspective. My conceptual thinking about the brain and the human condition began to change in 1994 as I grew tired of disease and the rather negative view of the clinical model and I began to understand that most brains are not diseased. The issue for me was and remains how we develop an approach that underscores health promotion and development of human potential.

The mid-1990s represented a major shift in my work as I focused on applying my knowledge of the human brain and brain function to health. I began to make presentations and publish papers on the role of learning and how the act of learning may be a health-promoting behavior (see Appendix 1 and www.paulnussbaum.com). I used relatively provocative terminology writing about a "learning vaccine," "schools as brain health centers," and "the role of education as a health concern." The seeds of brain health began to emerge and my early thoughts were formed with the understanding that learning served a primary function for the brain, almost a life sustaining nutrient for the central nervous system. The challenge was that while I believed the brain was the single greatest system ever designed, most of the nation, including the health delivery system, did not give it much attention or priority. How could I and others encourage America to champion brain health?

Converging Forces

To my great good fortune, several major conceptual shifts emerged in the 1990s that increased in momentum with the onset of the 21st century. The United States and much of the world is undergoing a demographic boom of older adults. Aging in the latter lifespan, therefore, is viewed as a much more important issue. The baby boom generation, approximately 76 million Americans born between 1946 and 1964, have attracted attention as a major force in our nation. Perhaps because of the boomers, health is now a focus of personal popularity and proactive behavior. The human brain is finally on the social "radar screen" and with it the birth of new enterprises, businesses, and interest in this important part of our being. Together, the aging population, baby boomer momentum, health, and new attention to the human brain have reinforced that the health of the human brain is the next great frontier for exploration, innovation, and application.

These forces make it conducive to educate the nation about the basics of the brain and the research-based activities that reduce their risk of dementia (See Nussbaum, 2007). My emphasis in this effort has been to speak and write for laypersons rather than as an academic for other academics. This is particularly important because it recognizes the fact that millions of Americans can benefit from brain health if they are provided information they can personalize. My goal is to make brain health fun, personal, and informative. If people can personalize something they are much more likely to change their behavior and be motivated to achieve a specific goal.

Brain Health Lifestyle

Brain Health and Wellness (Nussbaum, 2003) was a first attempt to consolidate the existing research on lifestyle and brain health promotion. This text included a primer on the basics of the human brain to underscore that the reader needs to understand the basics of the brain to make behavioral changes. The personalization of brain health emerges naturally from the fact that millions of Americans, particularly baby boomers, are serving the role of caregiver for a parent who has a dementia such as AD. Policy makers need to address the epidemic of dementia in our nation before we are completely overwhelmed emotionally and economically. According to surveys, Americans fear losing their memory and we are in need of something to help offset this fear with a proactive and hopeful direction for everyone.

A short guide entitled *Love Your Brain* (Nussbaum, 2005) was written in partnership with MetLife Mature Market Research to offer the consumer specific activities to promote brain health at almost any age. The developmental lifespan approach to brain health was quite popular and recognized that the brain at any age requires healthy input for both short term and long-term benefit. Section I of this book, originally published separately as *Your Brain Health Lifestyle: A Proactive Program to Preserve Your Lifestory*, provided my most comprehensive and integrated lifestyle program for those interested in brain health promotion. Like the other two books, this text first presented the basics of the brain and then described in layperson terms how a certain lifestyle might promote brain health by building "brain reserve."

My brain health lifestyle program is based on five major areas of lifestyle: physical activity, mental stimulation, nutrition, spirituality, and socialization. These five areas were derived from animal research in the 1950s that demonstrated environmental input affects the brain and brain function. Animals raised in an enriched environment demonstrated larger brains, new brain cells in the hippocampus (a structure critical for learning and memory) and more cellular connections (brain reserve) relative to animals raised in unenriched environments. My review of this work found that scientists who published this research defined enriched environments as having three major factors:

Integration of Rigor and Relevance with the Human Brain

1. Socialization: having other animals in the environment to interact with
2. Physical activity: having a running wheel on which the animals could run
3. Mental stimulation: having toys in the environment to play with

These studies were convincing and there is no doubt that animal brains demonstrate a positive or negative effect from environmental input.

Humans and the Enriched Environment

There was much debate into the 1990s whether the human brain actually was affected by environmental input. In 1998, however, a significant study was published in *Nature* (Ericsson et al., 2003) that indicated the human brain had the ability to generate new brain cells (neurogenesis) identical to that of the rodent as established in the animal research. Interestingly, and certainly not coincidentally, the hippocampus was the area of the human brain found to generate new brain cells. This finding challenged many years of traditional clinical scientific thought that the human brain was a rigid and degenerative system not capable of new brain cell development. The idea of a "critical period of brain development" underscored the traditional thinking. Neurogenesis is one product of a brain that is highly dynamic, constantly reorganizing, and malleable. The human brain is thus thought to have "plasticity."

Following findings from the animal research, a basic and perhaps obvious question is what leads to new brain cell development in humans? The answer appears to be the same as found in the animal studies. Research in humans has found that socialization, physical activity, and mental stimulation are very important for reducing the risk of dementia. Two other factors, nutrition and spirituality have also been found to have important positive effects on the brain. These five factors represent the foundation for my brain health lifestyle that links back to the animal studies of the 1950s and the relationship of environment to brain structure and function.

Becoming Acquainted with the International Center for Leadership in Education

I had the distinct pleasure of meeting Joe Shannon, now Deputy Chief Education Officer at the International Center, in 2006 when I was delivering a talk on brain health and education. He told me about the important work being done at the International Center and arranged for me to deliver a talk on brain health and education on July 4, 2007, in Washington, D.C. Interestingly, although it was the Post-Conference and a national holiday in our nation's capital, there were more than 1,000 people in attendance to hear this message. Since that time, I have learned more about the work of the International Center, the dedicated staff, and the president, Dr. Bill Daggett. It became clear that the International Center understood the important need for applied brain research and education in the nation. The International Center is a leader in understanding the education system in the nation, the cultural shifts that are needed to propel learning and teaching in the 21st century, and the need for innovation in the education system. Together we have the opportunity to build upon the exciting concepts of brain and education that I discussed in the 1990s and to create new avenues

Integration of Rigor and Relevance with the Human Brain

for education-based application of what promotes brain health and learning/teaching.

The Interface of Rigor, Relevance, and the Human Brain

I have studied the rigor, relevance, and relationships approach espoused by the International Center for two reasons. First, I simply wish to learn about the model as a person interested in education and second, I want to think about and discuss the interface of the model to human brain processing. More specifically, can we explain the underlying brain mechanisms of the Rigor/Relevance Framework and does this understanding permit us to refine and enhance both learning and teaching approaches?

Basics of the Human Brain

The human brain is indeed the most brilliant system ever designed! Weighing between two and four pounds and comprised of millions of brain cells (neurons), the brain provides our every thought, emotion, and behavior. We refer to the brain as the "central nervous system" and it is our information processing system, encoding, filtering, and acting upon thousands of bits of information daily. The brain facilitates our endless life experience into a wonderful autobiography I refer to as your "life story." It is your life story that needs protecting so that you may share it with your next generation. It is also this life story that becomes eroded with cruel diseases of the brain such as AD.

Your brain can be organized several ways for the purposes of understanding. First, the brain has a cortex and a subcortex. Cortex

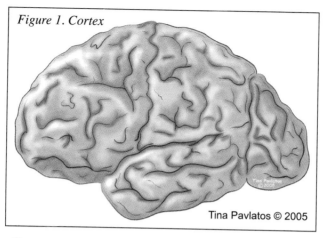

Figure 1. Cortex

literally means "bark of a tree" and it is the outer, convoluted part of the brain (see Figure 1). The cortex is a conscious processor of complex information and permits you to plan, remember, learn, use language, execute behavior, and appreciate space around you. At present, we probably significantly underestimate the power and ability of the cortex and I believe future generations will learn how to better unleash this miracle system.

The subcortex (see Figure 2) is a more primitive and older part of your brain. The name refers to the fact that this part of your brain sits just under the cortex. The subcortex is responsible for procedures, rote skills, and more automatic responses. It tends to function more on a subconscious level relative to the cortex. Important to the brilliance of the brain architecture, the cortex and subcortex are connected and do not exist in isolation. There are several critical pathways that link the subcortex to the cortex and permit ongoing information sharing and coordination of action. Disease or injury to one or the other affects both because of the connection.

Integration of Rigor and Relevance with the Human Brain

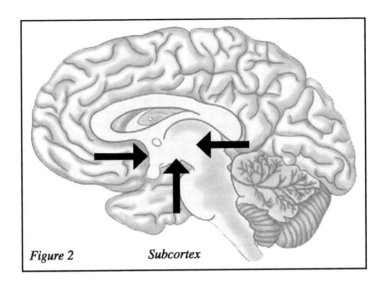

Figure 2   **Subcortex**

A second organizational schema for understanding your brain is a "hemisphere approach." The brain has two hemispheres or sides and we label them simply as left and right (see Figure 3). Each hemisphere is connected by a bridge of cells referred to as "white matter" known as the corpus collosum. The corpus collosum is literally a bridge that connects the two sides of your brain and permits information flow between the two sides. Interestingly, it is thought that the female brain has a larger corpus collosum and that the female brain uses both sides for processing more efficiently than male brains that tend to be more unilateral processing systems.

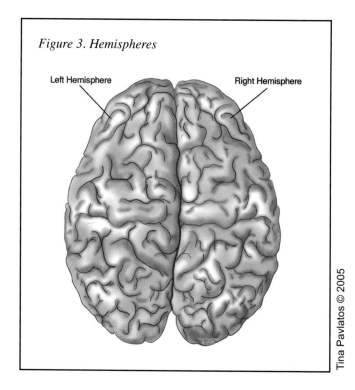

Figure 3. Hemispheres

For most people the left hemisphere is a primary language area and processes verbal information, academics, tasks, details, and sequences, and it is analytic. Western civilization and our current education system rely heavily on the left hemisphere. We refer to the language hemisphere as "dominant" because we have such respect for language. The right or "non-dominant" hemisphere relies much less on language and instead processes information such as faces, size, shape, space, and tone or melody. Our culture has not tapped the non-dominant hemisphere yet and this is a fertile area for development of human potential.

Integration of Rigor and Relevance with the Human Brain

A third major approach to understanding your brain is to recognize the four major "lobes" or regions of your cortex (see Figure 4). Each hemisphere has a frontal lobe, temporal lobe, parietal lobe, and occipital lobe. These regions interacts, but they have individual specificity to help process information in distinct ways.

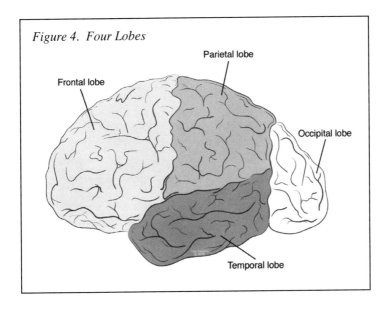

Figure 4. Four Lobes

Frontal Lobes. The frontal lobe is the youngest of the four lobes having evolved most recently. Perhaps the most sophisticated of the four lobes, the most mysterious, and the most important for "Quadrant D" as the frontal lobe is a problem solver, particularly in novel, complex, and unpredictable worlds. Also known as the "executive system" the frontal lobe is a type of CEO of the brain. While it does not have a direct impact on IQ, it does help to choreograph and orchestrate the behaviors of the rest of the brain. Your frontal lobe helps you:

- plan
- strategize
- organize
- structure
- execute behavior
- routinize
- time estimate
- conceptualize
- categorize
- mentally shift
- control mood
- maintain attention
- control impulses
- sequence
- express speech
- control motor output

The frontal lobe is very important to daily life and your personality, and without a healthy frontal lobe a person will have great difficulty socially, occupationally, and with interpersonal situations. Many of the psychiatric illnesses are related to a structural or neurochemical breakdown within the frontal lobes.

Temporal Lobes. Another critical region of the cortex is the temporal lobe. Sitting just under each temple on the side of your skull is a region important to auditory function, language comprehension, appreciation of a higher being and spirituality, memory, new learning, and spatial representation of the world. The hippocampus sits in the middle region of each temporal lobe (See Figure 5), helping to encode new information and serving as the entry to new learning. It represents one of the most important structures in our body and serves as the foundation for learning, the education system, and the compilation of our life story. Without a healthy hippocampus we would be rendered dependent.

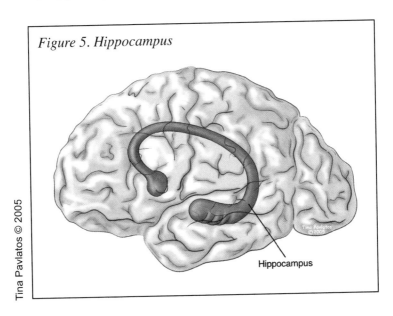

Figure 5. Hippocampus

As noted earlier, the animal and human hippocampi have the unique ability to generate new brain cells. It is thought that an enriched environment filled with the novel and complex promotes the health of the brain and the hippocampus. The brain health lifestyle described earlier is a guide to the enriched environment, the novel and complex, and to brain reserve. In contrast, studies indicate placement of the hippocampus in an unenriched environment or chronically stressful environment can lead to structural and functional damage to the hippocampus and long-term negative consequences such as a higher risk for AD. This is supported by research that finds a lower IQ and being raised in poverty in childhood increases the risk of AD later in life. It is important to recognize that early environments such as the school setting have both short-term and potentially long-term consequences. The purpose of this text and the work of the International Center with Dr. Nussbaum is to present the health-promoting pathway.

Parietal Lobes. The parietal lobes sit towards the top and back of the cortex just on top of the temporal lobes and behind the frontal lobes. This is a complex region of the brain that serves as a type of "town forum" or "association region" where input is received from the occipital and temporal lobes to form an integration of knowledge. Behaviors such as transcribing what is heard into written prose or reading and comprehending are examples of such knowledge integration that requires multiple inputs. Your parietal lobes also are important for arithmetic ability, visuospatial skill (finding your car in the parking garage!), visuoconstruction (such as building a puzzle), and even some short-term memory ability.

Occipital Lobes. The occipital lobes sit at the rear of the cortex and represent your visual system. Vision is a highly complex behavior and the occipital lobes can both perceive and distinguish what is seen. The left occipital lobe processes visual information from the right visual field while the right occipital lobe does just the opposite. Obviously your occipital lobes are critical for facial discrimination and recognition, reading, and appreciation of different signs and shapes.

Integration of Rigor and Relevance with the Human Brain

Your Brain as a Processing System

So far we have reviewed the importance of environment to brain structure and function, your brain as having plasticity and an ability to generate new brain cells toward brain reserve, and the structural makeup of the brain. It is important to now focus on the action of your brain and how this small miracle actually processes information more efficiently than anything on the planet! Although we do not know all the answers about how the brain processes information, I will proceed to describe in basic terms what we think we know about this area.

Primary, Secondary, and Tertiary Levels of Processing

You are well aware that you have five major sensory systems: (1) vision, (2) audition, (3) tactile, (4) olfactory, and (5) gustatory. The brain processes information at a basic level using each of these sensory pathways. Your experience of the world would change dramatically if you lost even one sensory system. The brain also appreciates when you stimulate it using multiple sensory pathways simultaneously! When your brain registers an incoming stimulus (visual object), regardless of the sensory pathway used, there is a basic level of processing referred to as primary level of processing. As this information is processed at a deeper level by the brain there is identification and recognition of the actual stimulus (animal vs. fruit), referred to as secondary level of processing. The brain then is capable of processing at an even deeper level using the same stimulus to interpret and prepare an action regarding the stimulus (the animal is a threat) by rapidly comparing what is recognized against past experience or existing knowledge stored in your brain. This latter and most deep level of processing is called tertiary processing.

Each of the five sensory pathways likely have this three-level processing approach, but Figure 6 highlights where the different levels of visual and auditory processing are thought to sit within

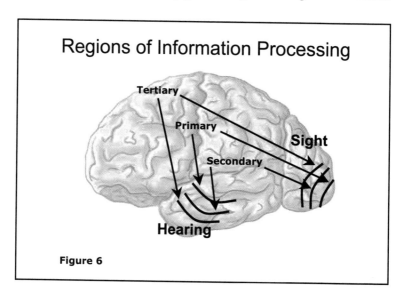

Figure 6

the occipital and temporal lobes, respectively. As noted earlier, the brain has the ability to process distinct sensory information, but also to integrate different types of information into an association at an "association region." Each lobe likely has an association region where the brain receives and integrates the distinct sensory information forming a higher, more complicated level of knowledge (see Figure 7).

Integration of Rigor and Relevance with the Human Brain

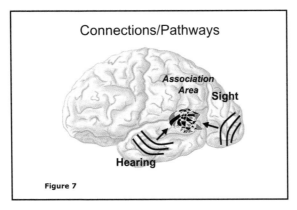
Figure 7

One of the major association regions is located in the parietal area of the cortex. Multiple pathways connect this association region to the frontal lobe where it is believed application of this integrated and complex information occurs (see Figure 8). The frontal lobe is the region of the cortex that applies or helps execute behavior.

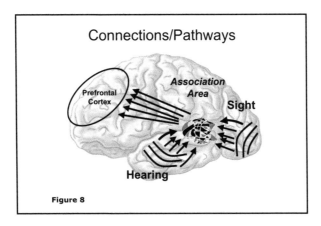
Figure 8

In summary, the brain receives information from one of the five major sensory systems. This registration of basic incoming information is transitioned to a secondary level of processing where there is some identification and recognition of the stimulus. The information is then processed more deeply at the tertiary level where the information is interpreted. Such deeply processed information is then potentially integrated with other incoming sensory information into an association region to form a more complicated, higher order level of knowledge. The association regions have multiple pathways to the frontal lobe where application or execution of behavior based on the complex knowledge is potentially performed.

Brain Processing and Rigor, Relevance, and Relationships

Using the previous discussion on brain processing we can now turn to an exploration of how the Rigor/Relevance Framework interfaces with such brain processing.

The Rigor/Relevance Framework is based on two dimensions of higher standards and student achievement. These include a continuum of knowledge that describes the increasingly complex ways the brain processes. *The Knowledge Taxonomy* is based on the six levels of Bloom's Taxonomy:

1. awareness
2. comprehension
3. application
4. analysis
5. synthesis
6. evaluation

Integration of Rigor and Relevance with the Human Brain

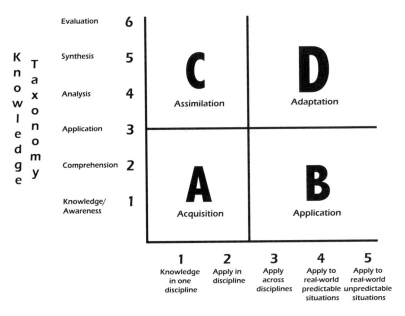

Application Model

According to this knowledge taxonomy the low end of the continuum involves acquiring knowledge and being able to recall or locate that knowledge in a simple manner. This is most similar to and likely related to brain processing at the primary or sensory level to secondary level of processing. The high end of the Knowledge Taxonomy labels more complex ways people use knowledge. At this level, knowledge is fully integrated by the brain and the person can do much more than locate information. Several pieces of information can be combined in both logical and creative ways as a type of assimilation or intellectual alloy. This is a higher-order level of thinking that assists one with problem solving using multiple approaches. It is this latter type of higher level thinking that is most consistent with information integration that occurs in association regions within the brain.

The second dimension of higher standards or continuum of knowledge was created by Dr. Daggett and is known as the *Application Model*. This model describes putting knowledge to use on a continuum from the low end where knowledge is acquired for its own sake to the high end where application occurs to solve complex real-world problems. There are five levels to the application continuum:

1. knowledge in one discipline
2. apply in discipline
3. apply across disciplines
4. apply to real-world predictable situations
5. apply to real-world unpredictable situations

The knowledge and application models are directly related to the information processing model of the brain described above (see Table 1). The brain processes information from a simple to more complex level, moving from sensory registration to integrated knowledge bases with a final outcome of behavior execution on the new, complex information. The most highly complex and efficient use of integrated information is in the form of application to unpredictable situations, particularly in rapid time. In other words, can the brain take the integrated information from the association regions and apply it to solve problems that are perhaps both complex and unexpected?

Integration of Rigor and Relevance with the Human Brain

Table 1: Knowledge and Application Models with Brain Processing Model		
Knowledge Taxonomy	**Application Model**	**Brain Level of Processing**
Awareness	Knowledge in one discipline	Primary and secondary levels
Synthesis	Apply across disciplines	Association areas of cortex
Evaluation	Apply to real-world unpredictable situations	Multiple pathway connections: Association regions to frontal lobe

Pursuing the interface of rigor and relevance to brain processing, one can begin to align the Rigor/Relevance Framework to both brain region and process. Quadrant A represents simple recall and basic understanding of knowledge for its own sake. This type of process is most aligned with primary and perhaps secondary levels of information processing from the brain perspective. Quadrant C embraces higher levels of knowledge and is most related to secondary levels of processing that are more deeply processed into the association regions of the cortex. Quadrants B and D represent action or high degrees of application, with Quadrant D being a more sophisticated level of application using multiple knowledge domains. Both Quadrants B and D are most related to the cortical processing that occurs when the integrated information created within the association regions is delivered to the frontal lobe. The application or action on such knowledge is the role of the frontal lobe or executive system. The difference between B and D from a brain processing perspective may involve level of sophistication, with Quadrant D involving more pathways of information flow from association areas to the frontal lobe, more complicated information, and more unpredictability regarding the information relative to Quadrant B. From a learning and teaching perspective it is Quadrant D that will require and stimulate the most sophisticated regions and processes of the human brain.

Summary

The explanation of how rigor and relevance of learning and teaching is applied to brain processing is a new and exciting area for discussion and thought. This section has proposed an interface of the two learning dimensions (Knowledge Taxonomy and Application Model) to the information processing system of the brain. By considering a direct interface between these dimensions to an underlying brain processing system, we can better understand how the brain processes or learns, and develop more enlightened models of teaching and learning. The most salient challenge for our education system appears to be the generation of models of teaching and learning that promote a high frequency of association area activity (mass integration of different forms of knowledge – Quadrant C) and provide opportunities for the student to apply the integrated knowledge in environments that offer unpredictable challenges and demand unpredictable solutions to problems (frontal lobe activity – Quadrant D). It is always a good idea to return to some basic truths when we explore and attempt to address these complex issues within the context of education:

Integration of Rigor and Relevance with the Human Brain

1. The human brain is a highly dynamic system that responds to environmental input. An environment that is enriched and offers complex and novel stimuli will promote brain health and advanced learning.

2. Learning is a fundamental health-promoting activity for the human brain as it builds brain reserve and stimulates the hippocampus, the critical structure for learning. This has both short- and long-term benefits.

3. The enriched environment includes socialization, physical activity, mental stimulation, spirituality, and nutrition. Schools have an opportunity to implement each of these brain health factors.

4. Education has both immediate and long-term effects, the latter being promotion of brain health and even reduction in the risk of AD.

5. The brain processes information beginning at the sensory level, moves to more complex levels, integrates forms of different information at the association regions of the brain, and eventually applies the integrated knowledge in the frontal lobes.

6. The Knowledge Taxonomy and Application Models are directly related to the information processing model of the human brain discussed in this section. Quadrant A is most

What Brain Research Teaches about RRR

aligned with the primary and secondary regions of the brain, Quadrant C is most aligned with the association regions of the brain, and Quadrants B and particularly D are most aligned with the frontal lobes.

7. Teaching, learning, and school environments can utilize this information to help develop more enlightened approaches for both short-term and long-term brain benefit.

Section III

Brain Research – The Basis of Successful Practices

Willard R. Daggett Ed.D.

Over the past couple of years at the International Center for Leadership in Education, we have gone across the United States to find the nation's both highest performing and most rapidly improving schools. We have had the privilege of working in cooperation with the Council of Chief State School Officers and with financial support from the Bill & Melinda Gates Foundation.

Several things became apparent as we located the highest-performing schools. First, while these schools did not know each other – in fact, they were not tied together formally or informally in any way – we found their practices to be strikingly similar in terms of what they taught, how they taught, and how they organized their schools. Second, these schools discovered their extraordinarily successful practices typically through trial and error. Finally, once they found their successful practices and began implementing them, they consistently told us that in retrospect, they realized what they did was return to common sense because their successful practices were based on common sense.

As we gathered and analyzed the nation's most successful instructional practices, we at the International Center also worked to understand the growing body of knowledge emerging from the explosion in brain research across the country. One of the primary reasons for the growth in brain research was the increased capacity of imaging technology such as CAT scans, PET scans, and MRIs. We began to compare what the schools were calling their common sense successful practices with the growing valued research on brain research. It became clear that while the successful schools had gone through trial and error to find their practices, their results corresponded to the results of sound brain research. We at the International Center concluded that educators need to acquaint themselves with the growing body of knowledge in brain research, and then translate that knowledge into education practices to be used in their classrooms.

Rigor, Relevance, and Relationships

The first of the successful practices that is consistent in brain research is the fact that relevance enables schools to achieve academic rigor with most students. That simple concept is what led to the International Center for Leadership in Education's Rigor/Relevance Framework, a tool that examines curriculum, instruction, and assessment.

Brain Research — The Basis of Successful Practices

Beginning from the day students enter kindergarten, K-12 schools face a daunting challenge in preparing those students for a changing world. Students will need the knowledge and skills that are necessary to compete in a technological and global economy. Beginning in elementary school, we need to raise the bar in providing a rigorous and relevant education to prepare all students for lifelong success.

For decades, the 3 Rs — 'reading, 'riting, and 'rithmetic — were considered the foundation of school instruction. Increasingly, though, as the challenge has gone out to K-12 schools to educate all students to high standards, the traditional 3 Rs are being replaced by a new set — rigor, relevance, and relationships.

Many teachers of young students, particularly those in kindergarten and 1st grade, are skilled in the relevance and relationship aspects. As humorously outlined in Robert Fulghum's popular book, *All I Really Need to Know I Learned in Kindergarten*, these youngsters learn the basic social skills that are relevant in everyday life at any age. Elementary students typically are taught in the context of real-world settings. They often work in teams, paralleling work in the real world. Hands-on projects encourage students to apply the concepts and knowledge they acquire, adding to their understanding of how to use what they are learning.

In the lower grades, students tend to stay with one or two teachers for the entire day, which promotes the development of strong relationships. Elementary teachers know their students well. Unfortunately, as students progress from elementary to high school, these personal relationships between student and teacher too often decline. The opportunities for application of instruction also tend to decline in the higher grades as teachers try to deal with curriculum overload and schools struggle to cope with new accountability requirements.

Because rigor is integrally connected with the other two Rs — relevance and relationships — performance tends to fade for some students in high school. Consider, for instance, that in math and science, U.S. 4th graders are among the top students in the world. By 8th grade, they are in the middle of the pack, and by 12th grade, they score near the bottom of all industrialized nations, according to the National Academies 2005 report, *Rising Above the Gathering Storm*.

Rigor/Relevance Framework

To assure the inclusion of both rigor and relevance, the International Center for Leadership in Education created the Rigor/Relevance Framework™. The Rigor/Relevance Framework is a tool developed by staff of the International Center to examine curriculum, instruction, and assessment. The Rigor/Relevance Framework is based on two dimensions of higher standards and student achievement.

First, there is a continuum of knowledge that describes the increasingly complex ways in which we think. The Knowledge Taxonomy is based on the six levels of Bloom's Taxonomy:

1. Awareness
2. Comprehension
3. Application

4. Analysis
5. Synthesis
6. Evaluation

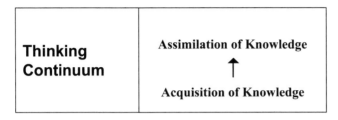

The low end of this continuum involves acquiring knowledge and being able to recall or locate that knowledge in a simple manner. Just as a computer completes a word search in a word processing program, a competent person at this level can scan through thousands of bits of information in the brain to locate that desired knowledge.

The high end of the Knowledge Taxonomy labels more complex ways in which individuals use knowledge. At this level, knowledge is fully integrated into one's mind, and individuals can do much more than locate information. They can take several pieces of knowledge and combine them in both logical and creative ways. Assimilation of knowledge is a good way to describe this high level of the thinking continuum. Assimilation is often referred to as a higher-order thinking skill; at this level, the student can solve multi-step problems and create unique work and solutions.

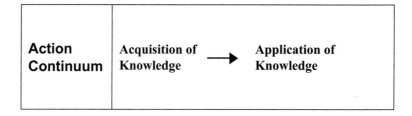

The second continuum, the Application Model, which I created and which provides the basis for much of the International Center's work, implies action. It has five levels:

1. knowledge
2. application of knowledge in a discipline
3. application of knowledge across disciplines
4. application of knowledge to solve real-world predictable problems or situations
5. application of knowledge to solve real-world unpredictable problems or situations

In the Application Model, knowledge is put to use. While the low end is knowledge acquired for its own sake, the high end signifies action – use of that knowledge to solve complex real-world problems, and to create projects, designs, and other works for use in real-world situations.

The Rigor/Relevance Framework has four quadrants. Quadrant A represents simple recall and basic understanding of knowledge for its own sake. Quadrant C represents more complex thinking as well as knowledge for its own sake. Examples of Quadrant A knowledge include knowing that the world is round and that Shakespeare wrote *Hamlet*.

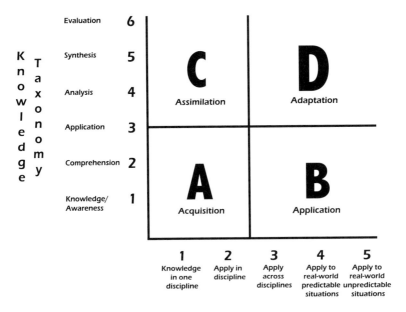

Application Model

Quadrant C embraces higher levels of knowledge, such as knowing how the U.S. political system works and analyzing the benefits and challenges of the cultural diversity of this nation versus other nations.

Quadrants B and D represent action, or high degrees of application. Quadrant B would include knowing how to use math skills to make purchases and count change. The ability to access information in wide-area network systems and the ability to gather knowledge from a variety of sources to solve a complex problem in the workplace are types of Quadrant D knowledge.

Each of these four quadrants can also be labeled with a term that characterizes the learning or student performance.

Quadrant A – Acquisition

Students gather and store bits of knowledge and information. Students are primarily expected to remember or understand this acquired knowledge.

Quadrant B – Application

Students use acquired knowledge to solve problems, design solutions, and complete work. The highest level of application is to apply appropriate knowledge to new and unpredictable situations.

Quadrant C – Assimilation

Students extend and refine their acquired knowledge to be able to use that knowledge automatically and routinely to analyze and solve problems and create unique solutions.

Quadrant D – Adaptation

Students have the competence to think in complex ways and also apply the knowledge and skills they have acquired. Even when confronted with perplexing unknowns, students are able to use extensive knowledge and skills to create solutions and take action that further develops their skills and knowledge.

Brain Research – The Basis of Successful Practices

The Rigor/Relevance Framework is a fresh approach to looking at curriculum standards and assessment. It is based on traditional elements of education yet encourages movement to application of knowledge instead of maintaining an exclusive focus on acquiring knowledge.

The Framework is easy to understand. With its simple, straightforward structure, it can serve as a bridge between school and the community. It offers a common language with which to express the notion of a more rigorous and relevant curriculum and encompasses much of what parents, business leaders, and community members want students to learn. The Framework is versatile; it can be used in the development of instruction and assessment. Likewise, teachers can use it to measure their progress in adding rigor and relevance to instruction and to select appropriate instructional strategies to meet learner needs and higher achievement goals.

Here is an example involving technical reading and writing.

Quadrant A
Recall definitions of various technical terms.

Quadrant B
Follow written directions to install new software on a computer.

Quadrant C
Compare and contrast several technical documents to evaluate purpose, audience, and clarity.

Quadrant D
Write procedures for installing and troubleshooting new software.

What Brain Research Teaches about RRR

Good instruction is not a choice of a single quadrant, but a balance. It may not be necessary for all students to achieve mastery in Quadrant A before proceeding to Quadrant B, for example. Some students may learn a concept better in Quadrant B when they see its application in a real-world situation.

The examples that follow illustrate different learning experiences in each of the four quadrants at the elementary and middle levels.

English - Elementary

RIGOR		Low RELEVANCE	High RELEVANCE
	High	C: Brainstorm as many words as possible to describe an object.	D: Create new words to describe phenomena or objects.
	Low	A: Memorize spelling words.	B: Write a story about a school experience using specific words.

Mathematics - Elementary

RIGOR		Low RELEVANCE	High RELEVANCE
	High	C: Find values in number sentences when represented by unknowns.	D: Develop formula for estimating a large quantity without counting (e.g., beans in a jar).
	Low	A: Memorize multiplication tables.	B: Collect outside temperatures for several days and graph the results.

Science - Elementary

RIGOR	Low Relevance	High Relevance
High	Make diagrams of animal life cycles. (C)	Design a zoo. (D)
Low	Make daily observations of an animal kept in class. (A)	Create a field book about organisms in a local river. (B)

Social Studies - Elementary

RIGOR	Low Relevance	High Relevance
High	Contrast citizens' responsibilities under different forms of government. (C)	Devise strategies for surviving a disaster (e.g., blizzard, tornado). (D)
Low	Memorize names, locations, and capital cities of the 50 states. (A)	Describe geographic and climatic characteristics of the local community. (B)

English - Middle Grades

RIGOR	Low Relevance	High Relevance
High	Compare and contrast the theme of two short stories. (C)	Write directions for assembling a product or carrying out a procedure. (D)
Low	Locate information in a technical manual. (A)	Assemble a product following written directions. (B)

What Brain Research Teaches about RRR

Mathematics- Middle Grades

		Low Relevance	High Relevance
RIGOR	High	**C**: Express probabilities as fractions, percents, or decimals.	**D**: Devise a scale to test consumer products and graph data.
RIGOR	Low	**A**: Plot the coordinates for quadrilaterals on a grid.	**B**: Make a scale drawing of the classroom.

Science- Middle Grades

		Low Relevance	High Relevance
RIGOR	High	**C**: Identify chemicals dissolved in an unknown solution.	**D**: Collect data and make recommendations to address an environmental problem.
RIGOR	Low	**A**: Construct models of molecules.	**B**: Collect data on dissolved oxygen, hardness, alkalinity, and temperature in a stream.

Social Studies- Middle Grades

		Low Relevance	High Relevance
RIGOR	High	**C**: Analyze primary and secondary documents.	**D**: Compare examples of stereotyping in historic and current events.
RIGOR	Low	**A**: View a historical video and answer factual questions.	**B**: Plan and participate in a community service activity.

Brain Research – The Basis of Successful Practices

No matter what the grade level, students require Quadrant B and D skills if they are to become lifelong learners, problem solvers, and decision makers. In essence, students need to *know what to do when they do not know what to do*. The Rigor/Relevance Framework provides a structure to enable schools to move all students toward that goal.

What was very apparent to me in the Application Model continuum was that schools with a traditional organizational structure of delivering disciplines (often in individual silos) were stuck on Levels 1 and 2. However our children need to be able to function at Levels 4 and 5 if they are to become independent in the world beyond school.

As my colleagues at the International Center for Leadership in Education and I began to use the Rigor/Relevance Framework in presentations and discussions around the country, it gained in popularity as a clear, comprehensive description of what schools need to do to prepare students for their future.

Young people need to be prepared for Quadrant D. Unfortunately, we find schools (especially at the high school level) typically teaching students in Quadrants B or C. Quadrant C represents the college-prep program, with strong academics but often little relevance to the curriculum. Quadrant B represents Career and Technical Education, for example, which has basic knowledge with a lot of real-world application but not especially sophisticated content.

Rigor and relevance is not a specific curriculum or instructional strategy. It is a way to evaluate and organize instruction and instructional practices. Schools that use rigorous academic instruction, but without a great deal of relevance, create a scenario of "teach it, test it, and lose it." In the world beyond school, what matters is not the exact subjects studied in school or even how students did on a test. What matters is how well they can apply that knowledge and skill to the world beyond school.

Interestingly, what we have found is when we teach students how to apply knowledge, especially in an area of interest to them, they can master academic rigor at a much quicker rate and retain it far longer than if the academic rigor is taught in isolation. I saw it as a classroom teacher, I saw it as a school administrator, I saw it with my own children, and now I see it in my own ten grandchildren. Relevance makes rigor possible.

Rigor, Relevance, and Brain Research

The relationship of the Rigor/Relevance Framework can be seen in how we process information. Figure 1 shows the brain. Our

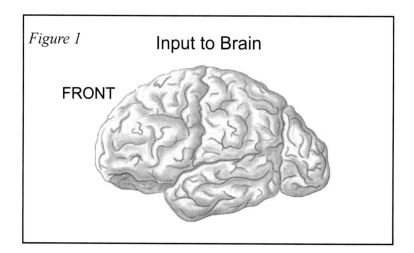

Figure 1 — Input to Brain — FRONT

eyes are in the front of the brain and, therefore, obviously where we see things. However, it is the back of the brain that processes sight. This simple overview will describe only the processing of sight. In reality, all five senses – sight, hearing, touch, taste, and feel – are all integrated together.

There are three zones for sight in the back of the brain, as shown in Figure 2. (Each of the five senses has three zones, as illustrated for hearing as well as sight in the illustration.) The three zones process increasingly sophisticated data.

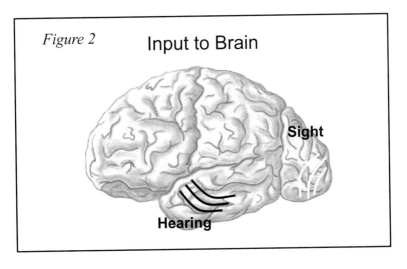

To illustrate, let's use a simple example of the family cat. In Zone 1 you see something and know it is an animal. In Zone 2 you know the animal is a cat. Now most people would ask, "Wouldn't everyone know that an animal is a cat?" The answer is, "No." A six-month-old child does not know the difference between a cat and a dog. A severely disabled adult might not know the

What Brain Research Teaches about RRR

difference. The human brain is very sophisticated. A computer would not know the difference between a cat and dog unless it had been specifically programmed to make that distinction.

In Zone 3 you see the cat and recognize that the cat is our family cat, Sylvester. Figure 3 is a picture of Sylvester. Sylvester adores my wife and follows her wherever she goes. She sits down, he jumps on her lap. If she picks him up, he puts her head on her shoulder and purrs. When she goes to bed at night, he runs up and jumps on the bed and puts his head next to her on the pillow.

Figure 3

Unfortunately, however, Sylvester does not like me. Figure 4 shows how he reacts as soon as he sees me – he is vicious! Sylvester also does not like my ten grandchildren. The pattern is getting more obvious. Sylvester is a one person cat.

One of our grandchildren, Erin, comes to our house quite often. She is extremely afraid of Sylvester. Can you blame her?

Figure 4

There is a second cat in the family named Ally, who is shown in Figure 5. Ally is a very peaceful, lovable animal. The grandchildren can use her as a pillow.

When Erin, the three-year-old granddaughter, comes to the house she is afraid of Sylvester and knows enough to stay away from him. She loves Ally and feels very comfortable playing with Ally. What does Erin do, however, when she encounters a third cat in a new location, for example at her cousin's house, that she has not been near before. Is the cat a good cat or a bad cat?

Figure 5

This simple example of evaluating an unknown cat illustrates where Quadrant A and Quadrant C in the Rigor/Relevance Framework fit in. Quadrant A is Zone 1, 2, and 3 (see Figure 6).

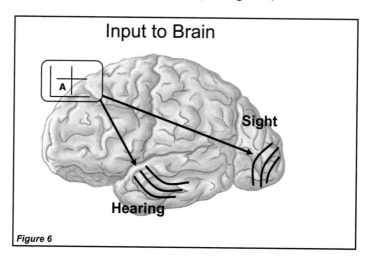
Figure 6

Quadrant C is what is known as the association area of the brain (see Figure 7).

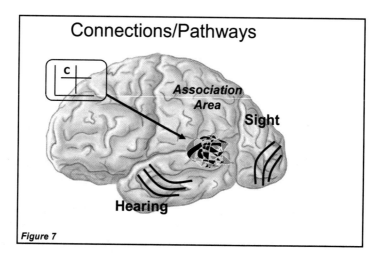
Figure 7

Brain Research – The Basis of Successful Practices

In the association area you analyze, synthesize, and evaluate information. You compare and contrast. That is the upper level of Bloom's Taxonomy or Quadrant C in the Rigor/Relevance Framework. To be able to use the association area of the brain or Quadrant C of the Rigor/Relevance Framework, you need preexisting knowledge. That preexisting knowledge comes from Quadrant A.

The front of the brain, called the prefrontal cortex, is where we encounter Quadrants B and D of the Rigor/Relevance Framework (see Figure 8). In the prefrontal cortex you take existing knowledge and experiences and apply them in new and innovative ways. Some people refer to this as thinking or, specifically, creative thinking. Some people refer to it as innovation, creativity, or design.

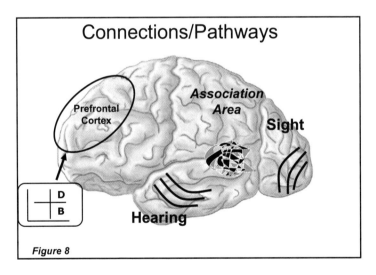

Figure 8

The use of Quadrants B and D (or the prefrontal cortex) is what will ultimately make an individual successful in the 21st century. Computers and technology are superb at Quadrant A and are also becoming better at Quadrant C. But computers cannot think, as artificial intelligence is still some years away. Therefore, the individual who can use the prefrontal cortex in areas of creativity, design, innovation, and creative thinking will be the person who has the greatest probability of success in the 21st century.

Successful Practices

Our analysis of the nation's most successful schools and most rapidly improving schools has taught us that teaching through application is a very effective way to engage students in pursuit of higher standards and to ensure that graduates can use what they have learned. When students see the relevance of what they are learning, they are motivated to learn more. Research has also shown that people retain more when they learn by doing as opposed to reading or listening. So why do teachers require students to read and listen so much? These strategies are appropriate in some learning situations, but surely not in all.

Every core subject has applications to the real world. Yet too often these subjects are reduced to textbook-driven memorization of facts. Science, for example, is about understanding the natural world. It seems logical that experiences in a science curriculum should give students direct opportunities to use science skills to make observations in and about the real world. Books are convenient for teachers but students need more.

Mathematics is a subject in which students develop the skills to recognize patterns, communicate relationships between quantities, and analyze data. Without real-world uses of these skills, they are solely intellectual pursuits. It is similar to learning the rules of the road and the skills for operating a car without ever actually getting behind the wheel and going for a drive.

English language arts is about learning means and forms of communication as well as understanding the culture through works of literature. Again, the real test of communication skills is the ability to use them effectively in actual situations. Yet neither speaking and listening skills, nor technical reading and writing, are typically emphasized in the English classroom.

Perhaps the most compelling research driving the need for a variety of instructional strategies is that most students learn best when instruction emphasizes application. To assist students in achieving high standards, the teacher must create learning environments that present students with challenging problems. The teacher must use instructional strategies that are aligned with students' learning styles, so that they can demonstrate knowledge and use their skills.

It is for this reason that the integration of the arts and academic education, the integration of career and technical education and academics, the integration of physical education and sports in academics, all enable students to do better in the academics. Unfortunately, what most schools do when students are struggling academically is remove them from the arts, career and technical education, sports, and physical education programs. Our evidence indicates that schools should do the opposite. Schools actually should move students into those programs, but only if those programs are willing to modify to become the B and D quadrants of the A and C academics that are being taught in the schools. The high-performing schools integrate programs together, which creates a new hybrid that does not yet exist in most schools.

CTE teachers have a unique opportunity to teach the essential English language arts, mathematics, and science skills while equipping students with the knowledge and skills needed for success in the workplace. The "hands-on" curriculum that characterizes career and technical education provides opportunities for students to apply the academic and technical skills that they have learned in the classroom in real-world applications on the job. No other curricular area lends itself more to real-world application than career and technical education.

Another example can be found in the arts. Both performing and visual arts are excellent ways to teach students the application of math, science, language arts, and social studies skills. This is especially effective when used with students who love the arts.

For students who like athletics and/or physical education, teachers can use it as the technique, the strategy, or the methodology to teach the application of math and science. Unfortunately, most schools are stuck in silos called disciplines. Disciplines are an effective way to arrange data and to teach content, but are very ineffective in teaching the application of knowledge and, therefore, learning.

Brain Research – The Basis of Successful Practices

To provide a structure to move to this integration of knowledge at the application level, many schools have abandoned the old position of department chairpersons. In their place, schools have created interdisciplinary chairpersons and given teachers common planning periods to plan their instruction and their student projects. The integration of academics into the arts or career technical education or physical education can take many forms. It can be as simple as an individual lesson.

It also can be as comprehensive as entire schools organized around that integration. The following are examples of such schools:

- **Saunders Trades and Technical High School** in Yonkers, New York, was the first public trades school in New York State, opening in 1909. There is a waiting list to enter the school. For more than 90 years, the school has been committed to combined vocational and academic instruction that provides graduating students the skills for employment and an academic preparation for college.

  Ninth grade students are assigned to small learning communities taught by core area teachers using interdisciplinary instruction and flexible scheduling to ensure a firm academic foundation. In addition, an exploratory career program facilitates the transition to high school and helps students develop strong study skills and a disciplined work ethic. For the following three high school years, students major in one of 13 technical, vocational, or occupational programs while continuing to be enrolled in college preparatory courses.

Graduation requirements exceed state requirements and those of other Yonkers high schools. To meet these requirements and a demanding schedule, students have a longer school day than students in the other four city high schools. Students and their parents choose Saunders because of its long-standing academic reputation, the integration of academic and employment knowledge and skills, and its reputation as a safe school in an urban setting.

- **Michael E. DeBakey High School for Health Professions** is an urban high school located in Houston that was formed as a partnership between the Houston Independent School District and Baylor College of Medicine. This small magnet school is an outstanding model of cultural diversity, community partnerships, and high expectations for all students.

  The school has a clear curriculum focus that supports its health professions theme. According to students, teachers, administrators, and parents, this focus is a major contributing factor to the school's success. Very high expectations are set for students. The staff stresses the importance of student achievement and devotes resources to ensuring that every student succeeds. Administrators, teachers, and students support each other and have created a culture of caring that serves as the foundation for academic success.

- **The Boston Arts Academy** is one of 13 pilot schools created by the Boston Public Schools in 1998. It is the first pilot school fully devoted to a program integrating an arts and academic curriculum. During its first six years of existence, it has become a flagship of education reform in grades 9-12 arts training. It models the concepts of success for all, diversity in its student body, and authentic training in the arts linked to a strong academic program. The faculty sought to develop a school that addresses the questions of what education means to students and the community, what success means in secondary education, and what arts education means in today's society. The mission of the school states that it is charged with being a laboratory and a beacon for artistic and academic innovation.

- **The Academy for the Arts, Science, and Technology** in Myrtle Beach, South Carolina, serves juniors and seniors from high schools throughout the Horry County School District. Students enroll in a major and take four to five academic courses offered on a year-round block schedule and also have the opportunity to enroll in high school concurrent college coursework at a local technical college and a local university. Academy coursework integrates academic and vocational instruction designed to prepare students for postsecondary education and their career choice.

- **The Toledo Technology Academy** is located in Toledo, Ohio. Nestled in an historic school, this innovative manufacturing engineering technology magnet high school enrolls 128 students in grades 9-12. Established in 1997, the school's strong performance is due to six distinct assets: the extraordinary partnership of the Toledo Public Schools with more than 60 companies in the area's manufacturing community; a rigorous, comprehensive academic and technical curriculum; small enrollment fostering small class sizes and collaborative hands-on learning opportunities; the highly qualified, committed professional staff; the school's involved and supportive parents; and a strong interdisciplinary curriculum. The mission of the school is captured in its slogan: "A different kind of school; a different way to learn" and in the motto of its students: "We conduct ourselves as if we are in a high tech, corporate business environment." The mission reflects best educational practices and reform initiatives resulting in significant improvement in student performance, such as the Baldrige in Education Initiative, High Schools That Work (HSTW), Project Lead The Way, Tech Prep; and the TTA founding principle of Total Quality Management (TQM). The philosophy of TQM promotes continuous improvement — a process for managing quality in everything. This philosophy mirrors the tenets of *No Child Left Behind* and Ohio's expectations of continuous improvement. It also incorporates all state academic and technical content standards.

Many other options exist for the integration of academics into other disciplines. These include schools within schools, charter schools, majors, etc. In all cases, the integrative leads to instruction in Quadrants B and D of the Rigor/Relevance Framework.

Successful Practices and Brain Research

Why are practices such as integrating the arts and sports into academics so successful? They are successful because they create more connections within students' brains – which helps them in all areas of their lives.

As Figure 9 shows, information flows from Zones 1, 2, and 3 of the associated sensory area of the brain to the prefrontal cortex.

Brain Research – The Basis of Successful Practices

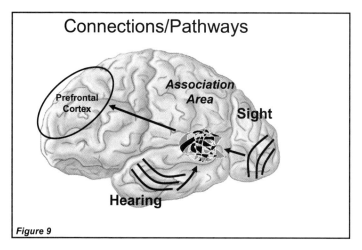

Figure 9

We know you need to create pathways for that information to flow. Schools create those pathways. However, have any of you ever had one of these experiences (especially those baby boomers who are still teaching in our schools): You go to make a phone call to a person who you have talked to a number of times in the past, but suddenly you cannot recall the phone number. You start driving to a location that you have been in the past, then cannot remember quite how to get there. Or perhaps you see someone you recognize, yet the name escapes you.

When you forget otherwise well-known information, a neuron may have broken off and blocked the pathway. The brain has tens of millions of neurons. During the first 20 years of life, those neurons explode in number. The growth of neurons begins to slow in your 20s, 30s, and 40s. As you get into your 50s and beyond, not only does the growth slow but some of the neurons begin to die – especially those not used often. That is why we say "if you don't use it, you lose it."

What Brain Research Teaches about RRR

As those neurons begin to break off (sometimes because of lack of use, sometimes by the aging process), they can block the pathway. Often that blockage is only temporary; you will soon recall the person's name or the phone number or the directions you could not remember. If it does not clear up, and if you have multiple occasions of this, it can be the early stages of dementia.

If you forget a person's name, you may try to remember his or her spouse's name, friend's name, or child's name. That may help you recall the person's name. Similarly, if you are driving and cannot recall the directions, then think of some related area and the turn you need to take may become obvious. What you just did is divert to a new pathway. When you have multiple pathways and one becomes blocked, you detour almost instantly to another pathway.

Why does teaching academics through the arts or CTE help students obtain knowledge that had been previously more difficult to obtain – and more importantly – help them retain that knowledge and be able to use it? We created multiple pathways within the brain. We can also simply call this "thinking."

The more pathways we create, the greater our ability to use and retain knowledge. Thus, the successful practices we find around the country are good not only for learning, but also and perhaps even more importantly, for retaining and being able to use knowledge throughout a lifetime.

Appendix One

Your Personal Brain Health Survey

Complete your Brain Health Survey to learn more about your own brain health!! Results are not scientific and are meant to help guide your brain health lifestyle. Your brain health profile is comprised of five major areas:

> Physical
> Mental
> Social
> Spiritual
> Nutrition

The following survey uses research-based information to propose a brain healthy lifestyle. The survey is to be completed prior to starting your brain health program (*Baseline Score*) and repeated every three months to document your progress.

What Brain Research Teaches about RRR

Do not be concerned if you score poorly at first. This is probably the first time you have considered your own brain health! You will notice improvement if you remain loyal to your brain health lifestyle.

PHYSICAL DOMAIN

Circle the response that best describes your behavior during the past three months.

<u>Score</u>

1.	I walk 10,000 steps daily.	5 points
	I walk between 5,000 and 10,000 steps daily.	3 points
	I do not walk.	0 points
2.	I engage in aerobic exercise three hours a week.	5 points
	I engage in aerobic exercise one hour a week.	3 points
	I do not engage in aerobic exercise.	0 points
3.	I garden more than one time a week during season.	5 points
	I garden one time a week during season.	3 points
	I do not garden.	0 points
4.	I dance more than one time a week.	5 points
	I dance one time a week.	3 points
	I do not dance.	0 points
5.	I knit more than one time a week.	5 points
	I knit one time a week.	3 points
	I do not knit.	0 points

Physical Domain Total Points _____/25

Brain Research – The Basis of Successful Practices

MENTAL STIMULATION DOMAIN

Circle the response that best describes your behavior during the past three months.

<u>Score</u>

1. I read more than the news on a daily basis. — 5 points
   I read one new book a month. — 3 points
   I do not read. — 0 points

2. I am fluent in more than one language. — 5 points
   I am learning a new language (including American Sign Language). — 3 points
   I am not learning a new language. — 0 points

3. I handwrite on a daily basis. — 5 points
   I handwrite once a week. — 3 points
   I do not handwrite. — 0 points

4. I travel to new places one time a week. — 5 points
   I travel to new places one time a month. — 3 points
   I do not travel to new places. — 0 points

5. I play a musical instrument. — 5 points
   I am learning to play a new musical instrument. — 3 points
   I do not play a musical instrument. — 0 points

6. I listen to classical music on a daily basis. — 5 points
   I listen to classical music once a week. — 3 points
   I do not listen to classical music. — 0 points

7. | I play board games or other cognitive games daily. | 5 points
I play board games or other cognitive games once weekly. | 3 points
I do not play board games or cognitive games. | 0 points

Mental Stimulation Total Points _____/35

SOCIAL DOMAIN

Circle the response that best describes your behavior during the past three months.

<u>Score</u>

1. I eat one meal with my family/friends every day. — 5 points
   I eat one meal with my family/friends weekly. — 3 points
   I do not eat meals with anyone. — 0 points

2. I have joined two or more new groups this year. — 5 points
   I have joined one new group this year. — 3 points
   I have not joined any new group the past year. — 0 points

3. I have started more than one hobby in the past year. — 5 points
   I have started one new hobby in the past year. — 3 points
   I have not started a new hobby in the past year. — 0 points

4. I speak to family or friends every day. — 5 points
   I speak to family or friends three times a week. — 3 points
   I speak to family or friends less than once weekly. — 0 points

5.	I engage in personally meaningful activity daily.	5 points
	I engage in personally meaningful activity once a week.	3 points
	I do not engage in any personally meaningful activity.	0 points

Social Domain Total Points _____/25

SPIRITUAL DOMAIN

Circle the response that best describes your behavior during the past three months.

<u>Score</u>

1.	I pray on a daily basis.	5 points
	I pray one time a week.	3 points
	I do not pray.	0 points
2.	I meditate on a daily basis.	5 points
	I meditate one time a week.	3 points
	I do not meditate.	0 points
3.	I engage in relaxation procedures daily.	5 points
	I engage in relaxation procedures one time a week.	3 points
	I do not engage in relaxation procedures.	0 points
4.	I get enough sleep daily to feel rested and energetic.	5 points
	I get enough sleep daily to feel somewhat rested and energetic.	3 points
	I do not sleep enough to feel rested or energetic.	0 points

5. I attend a formalized place of worship weekly. 5 points
   I attend a formalized place of worship monthly. 3 points
   I do not attend a formalized place of worship. 0 points

Spiritual Domain Total Points _____/25

DIET AND NUTRITION

Circle the response that best describes your behavior during the past three months.

<u>Score</u>

1. I eat several ounces of salmon (or other fish listed in this text) two or more times a week. 5 points
   I eat salmon (or other fish listed in this text) one time a week. 3 points
   I do not eat fish. 0 points

2. I eat 2 cups of vegetables and fruits every day. 5 points
   I eat 2 cups of vegetables and fruits once a week. 3 points
   I do not eat vegetables and fruits. 0 points

3. I drink one 4 to 6 ounce glass of red wine or grape juice daily. 5 points
   I drink one 4 to 6 ounce glass of red wine or grape juice weekly. 3 points
   I do not consume red wine or grape juice. 0 points

4.	I eat two meals with utensils daily.	5 points
	I eat one meal with utensils daily.	3 points
	I do not use utensils on a daily basis.	0 points
5.	I eat walnuts several times a week.	5 points
	I eat walnuts once or twice a month.	3 points
	I do not eat walnuts.	0 points
6.	As a rule, I consume 80% of the portions provided to me.	5 points
	I consume 100% of the food on my plate.	3 points
	I tend to overeat regardless of portion size.	0 points

Diet and Nutrition Total Points _____/30

Date	Physical	Mental	Social	Spiritual	Diet
Baseline Score	___/25 = X 100 =	___/35 = X 100 =	___/25 = X 100 =	___/25 = X 100 =	___/30 = X 100 =

What Brain Research Teaches about RRR

YOUR BRAIN HEALTH SCOREBOARD

1st Quarter	___/25 = X 100 =	___/35 = X 100 =	___/25 = X 100 =	___/25 = X 100 =	___/30 = X 100 =
2nd Quarter	___/25 = X 100 =	___/35 = X 100 =	___/25 = X 100 =	___/25 = X 100 =	___/30 = X 100 =
3rd Quarter	___/25 = X 100 =	___/35 = X 100 =	___/25 = X 100 =	___/25 = X 100 =	___/30 = X 100 =
4th Quarter	___/25 = X 100 =	___/35 = X 100 =	___/25 = X 100 =	___/25 = X 100 =	___/30 = X 100 =
Annual Brain Health Score	___/100 = X 100 =	___/140 = X 100 =	___/100 = X 100 =	___/100 = X 100 =	___/120 = X 100 =

How to Calculate and Interpret Your Scores

1. To derive your *Quarterly Brain Health Score for each Domain*: Add the scores of your circled responses and insert the total score into the formula listed for each quarter.

2. To derive your *Quarterly Total Brain Health Score*: Add the total scores for each domain, divide by 140, and then multiply by 100.

3. To derive an *Annual Brain Health Score by Domain:* Add the four scores of each domain and apply that score to the formula listed for the Annual Brain Health Score for that particular domain.

    For example: Physical Health Domain:

    Total score for each quarter/100 X 100 = _____

4. To derive a *Grand Total for Overall Brain Health for the Year*: Add the five Annual Brain Health Scores together, divide by 560 and multiply by 100 *Grand Total* = (five Annual Brain Health Scores) /560 = _____
    X 100 = _____

What Brain Research Teaches about RRR

Interpretation of Your Score

You may use the following guide to interpret each of your four scores previously described:

100-90: Great job! Maintain your lifestyle approach.

89-80:	Good job! Make a few changes to improve your lifestyle.
79-70:	Average job. Consider making changes in several domains.
69-60:	Poor job. Significant change is needed in several domains.
59-50:	Help! Re-assess the importance of your life story and attempt to make one small change in your lifestyle at a time.

Maintain the Checkup for Your Neck Up!

Appendix Two

INTRODUCTION TO FOOD AND YOUR BRAIN

Lauren M. Dorgant

With all the mixed advice about diets, nutrition, and your health — who knows who or *what* to believe! There are TV commercials, magazine ads, the infamous late night infomercials on a "new and revolutionary" diet pill, or a "recently discovered" magic herb in the Swiss Alps. Then there are your friends, family members, and coworkers who always seem most convincing. There is literally diet advice around every corner. Who has the best recommendation? What *are* the healthiest ways to diet? What will keep you and your body healthy for life?

Eating a well-balanced, healthy diet can help you fit into your favorite pair of jeans, and can also help you reduce your risk of chronic age-related brain diseases such as dementia. Current research in diet and brain health has shown that nutrition plays a

What Brain Research Teaches about RRR

large part in brain development and function throughout the life cycle. There are changes you can make *now* to prolong the health of your body's most vital system, your brain.

Before starting any diet or taking any nutrition advice you should ask yourself a simple question, "Is this a plan I can continue to follow one day from now, one week from now, and even one year from now?" If you cannot truthfully and realistically answer "yes" to these questions, then the diet is probably not very practical for you. A healthy lifestyle requires you to make changes for the benefit of your overall health. You should not think of these changes as losses, but rather as steps you can take now to improve your overall current health, reduce your risk of future ailments, and generally improve your day-to-day mental health and well-being.

Your brain is a complex system that requires vitamins, minerals, and antioxidants for daily functioning and mental sharpness. The good news is that you can find these powerful vitamins, minerals, and antioxidants right in the foods you eat.

There are several components of a healthy diet that aid in promoting overall brain health and function. Some of the most essential include a group of *antioxidants* like carotenoids, vitamins A, C, and E; the mineral selenium; and folate. Other important components that can help combat mental health decline are the phytochemicals and omega-3 fatty acids.

You have probably heard that a diet rich in antioxidants can help prevent minor illnesses like colds, and reduce the risk of major illnesses such as heart disease. Antioxidants may also can keep the brain operating at its peak to help reduce the risk of serious conditions such as AD. Antioxidants are powerful substances that help counteract the damaging effects of oxygen in body tissues. Antioxidants play a housekeeper's role by essentially "cleaning up" free radicals before they can cause damage to your body. Free radicals are highly reactive atoms formed when oxygen and certain molecules interact. These free radicals have an unpaired electron that can cause cell damage, eventually leading to the decline in function or even cell death. Antioxidants are the main defense against these damaging free radicals.

Carotenoids are a class of pigments that give many fruits and vegetables their color and can range from yellow to orange to red or purple. Without carotenoids, for example, carrots would not be orange and tomatoes would not be red. Carotenoids function as antioxidants by helping to protect plants from harmful free radical damage. In the human body, carotenoids work to guard cells from the danger of free radicals that can be produced during metabolism from cigarette smoke and other pollutants or even from stress.

There are more than 600 known carotenoids, but one of the most recognized is beta carotene. Beta carotene is necessary for the creation of vitamin A by the body and is responsible for the pigment in deep orange-colored foods.

Some fruits and vegetables that are particularly rich sources of beta-carotene include:

- carrots
- pumpkin
- salmon

- sweet potatoes
- tomatoes
- cantaloupe
- peaches
- apricots

Vitamin A, also known as retinol, can be made by the body from beta-carotene. Vitamin A helps protect orange, yellow, and dark green vegetables and fruits from the sun's harmful radiation. It is also believed to have a comparable role in the human body. Vitamin A is essential for vision, cell growth, and healthy skin, teeth, and bones.

Good sources of vitamin A include:

- carrot juice
- sweet potatoes with peel
- pumpkin
- spinach
- collard greens
- kale
- beets

Vitamin C, also known as ascorbic acid, is another antioxidant that helps to protect from free radical damage. Getting enough vitamin C is a key to keeping your immune system healthy and protecting you from the common cold and other illnesses. Your body is incapable of producing this necessary vitamin so it is crucial that you acquire it daily from the diet. Vitamin C is also a primary ingredient in the formation of collagen, the glue that holds your cells together. It is used to form the skin layer that acts as a natural barrier during the healing process of wounds and cuts.

Excellent sources of vitamin C include the citrus fruits. Some examples are:

- oranges
- lemons
- limes
- grapefruit

Other good sources of vitamin C include:

- green peppers
- broccoli
- green leafy vegetables
- strawberries
- blueberries
- raw cabbage
- tomatoes

Vitamin E is used by the body to help protect the nervous system and the retinas of the eye, to help lower your risk for heart attacks and heart disease, and to protect you from irregular blood clotting that can prevent strokes. It may also be helpful in lowering your risk for certain kinds of cancer, particularly melanoma, because it works to stop skin cancer cell growth. Vitamin E also works with vitamin A to protect the lungs against pollutants in the air and to possibly lessen cellular aging, It also guards immune system function. More important, vitamin E can reduce your risk of neurodegenerative diseases like Alzheimer's by protecting the brain cells from damage.

Good sources of vitamin E include:

- nuts
- seeds
- whole grains
- green leafy vegetables
- wheat germ
- canola oil and other vegetable oils
- fish liver oil

The mineral selenium is another antioxidant that helps prevent free radical damage to protect your brain against unhealthy aging. It also can help to prevent dangerous blood clots including "brain attacks" (strokes). Selenium is found primarily in nuts and seeds, but can be found in other foods as well.

Good sources of selenium include:

- Brazil nuts
- canned tuna
- shellfish
- lean cuts of beef, turkey & chicken
- enriched noodles
- oatmeal and other grains
- cottage cheese
- garlic
- sesame seeds

Folate, also known as folic acid, is one of the B vitamins that have been proven to be beneficial for brain health. In one study, the risk of the development of Alzheimer's disease was reduced by 55% by eating the recommended daily allowance of folate. One explanation for this finding is that elevated homocysteine levels are known to cause damage to brain cells. Getting an adequate amount of folate helps to lower homocysteine levels in the blood that can in turn reduce the risk for AD.

Good sources of folate include:

- fortified breakfast cereals
- chickpeas
- asparagus
- spinach
- black, kidney, and lima beans
- Brussels sprouts
- oranges
- broccoli

Phytochemicals

A *phytochemical* is any nutrient or chemical derived from a plant source that is not considered to be a necessity for normal body functions, but can have beneficial effects on health. Phytochemicals are found primarily in fruits and vegetables and appear to promote health in several ways. They may slow the aging process and also reduce your risk for disease. Some of the phytochemicals beneficial to your health include lycopene, reservatrol, and the flavonoids.

Found in tomatoes and tomato products, pink grapefruit and watermelon, **lycopene** serves as an antioxidant that may aid in the reduction of heart disease and even the prevention of prostate cancer. Lycopene can also help to ward off the effects of free radicals on the brain.

Found in red grapes, wine and peanuts, reservatrol may help reduce your risk of heart disease and cancer. Because wine contains reservatrol, it is known to be beneficial for cardiac health when consumed in small amounts. The reservatrol in wine can also be beneficial for brain function by perhaps extending the life of the brain cells. This process occurs by stimulation of an enzyme, which interacts with proteins that regulate reactions to stressors and longevity. Although recommendations for wine intake are controversial, one glass (4-6 ounces) of wine a day for women and up to two glasses for men is generally considered sufficient. People who are predisposed to alcohol addiction or breast cancer and those told by their physicians not to consume alcohol should heed that advice.

Flavonoids are organic compounds found in fruits, vegetables, and some beverages that, according to recent studies, may be beneficial by acting as antioxidants. Certain flavonoids, like those found in beer, have been shown to have surprisingly powerful antioxidant effects; some exceed the flavonoids found in soy, tea, and red wine. Beer, like wine or any alcoholic beverages, should be consumed in moderation (see preceding advice). There are over 4,000 known flavonoids.

Some good sources of flavonoids include:

- vegetable and fruit skins
- soy
- tea
- citrus fruits
- onions
- hops and beer
- red wine

Anthocyanins are one of the most visible flavonoids. Anthocyanin pigments can range in color from deep red to purple or blue. This compound is responsible, for example, for making blueberries blue. Anthocyanins are found in flowers and fruits, but are also present in roots, leaves, seeds, stems, and other parts of plants. Together with the carotenoids, these two antioxidant-like compounds are responsible for the classic-colored tree leaves of the autumn season. Anthocyanins are also known for their antioxidantlike effects of cleaning up free radicals.

Some foods that contain anthocyanins include:

- blueberries
- cranberries
- strawberries
- blackberries
- cherries
- kiwi
- plums
- eggplant

An added benefit of these foods is that all can help to prevent urinary tract infections! Fresh berries may be expensive if out of season, so a comparable and healthy substitute is frozen berries.

The Truth About ... Dark Chocolate!

Along with being a very savory treat, dark chocolate also contains flavonoids. However, there are a few things you should know before you go on your next mid-afternoon chocolate binge! Chocolate is a food that is very high in calories, saturated fat, and sugar. Interestingly, dark chocolate may benefit your health, but only when *occasionally* eaten in *moderate* amounts. The benefit comes from the cocoa that is derived from the cacao plant, which is known to be rich in flavonoids. According to a recent study, dark chocolate rich in flavonoids may help prevent bad cholesterol (low-density lipoprotein or LDL) from causing plaque to buildup in the arteries of the heart. Another study showed that these flavonoids could possibly lower blood pressure in persons with hypertension. Both benefits could possibly help prevent strokes in persons with high cholesterol or high blood pressure. For the occasional dark chocolate indulgence, take your time and *enjoy* the creamy texture and rich flavor of the chocolate — you may realize that one square is just enough to satisfy your craving.

Which Foods Have the Most Antioxidant Power?

You may be interested in knowing what foods have the best antioxidant power for diet and health. In 2004, researchers at the U.S. Department of Agriculture (USDA) conducted their largest analysis of 100 commonly eaten foods that contain the largest concentrations of antioxidants per serving. The results of the study confirm the long-time belief that fruits and vegetables are an essential part of general health and wellness. The following are some of the foods that contain the highest levels of antioxidants according to the USDA research.

Fruits
cranberries
blueberries
blackberries
black plums

Nuts
pecans
walnuts
hazelnuts
pistachios

Vegetables
pinto and kidney beans
artichokes
black-eyed peas
red cabbage

Spices
ground cloves
ground cinnamon
dried oregano
turmeric

Omega-3 Fatty Acids

Fish is another food that can help to ensure healthy function of the cells of the brain because it is rich in a substance known as omega-3 fatty acids. Because our bodies do not naturally produce omega-3 fatty acids, we are reliant on our diet to provide them. Fish also does not have the high saturated fat content of other fatty meat products.

Fish especially rich in the omega-3 fatty acids include:

- mackerel
- lake trout
- herring

- sardines
- albacore tuna
- salmon

Recall that your brain is comprised of at least 60% fat and the fatty or lipid part of your brain helps to transmit information rapidly across your neural networks. Consumption of omega-3 fatty acids is thought to help maintain proper fat in your brain and to facilitate information processing. Indeed, some research demonstrates a relationship between several ounces of salmon a week and a reduction in the risk of dementia.

Other Lifestyle Factors: The Factors of Life

Other lifestyle factors such as obesity, high blood pressure or cholesterol levels, and inadequate exercise may also contribute to dementia.

Obesity

Being overweight or obese has been known to cause many health-related illnesses such as heart disease, diabetes, and high blood pressure, among countless others. Can being overweight or obese also lead to a decline in brain function? The answer is yes! Being overweight or obese during midlife has been shown to increase the risk for decline in brain function later in life, according to scientific research. For one study, researchers Whitmer and Yaffe discovered that, compared to people of normal weight, those who are overweight or obese in their 40s have a much greater risk of developing of AD.

This study included 9,000 participants ages 40-45 who were deemed overweight or obese by way of skin-fold fat measurements at the beginning of the trial. After follow-ups were conducted (up to 30 years for some participants), 221 cases of AD were diagnosed. Conclusions from the research show that participants in their 40s with the highest skin-fold fat measurements were about three times more likely to develop the disease than those with the smallest skin-fold fat measurements. The research also concluded that those who had the largest arm measurements were 2½ times more likely to develop AD than those in the group with the smallest skin-fold fat measurements.

This decade-long research-based evidence emphasizes that being overweight or obese plays a role in late-life brain functionality. Although there is no clear evidence of the exact correlation between obesity and dementia, researchers believe several mechanisms explain the link. Some of these mechanisms include fat regulation hormones leptin and adiponectin, co-morbid medical conditions, and inflammatory response proteins called cytokines. These results further stress the role of diet and nutrition as key factors in maintaining a healthy weight and promoting lifelong brain health.

Your Ideal Body Weight

Are you at your ideal body weight? Are you at risk for developing dementia later in life? An easy way to find your own ideal weight is through the following formula:

Ideal Body Weight (IBW) * =

If you are male:
106 lbs for 5 feet tall. Add 6 lbs for every inch over 5 feet.
Ex: Dustin is 5'10" tall.
Therefore his IBW is:
106 + (6 lbs x10")
106 + 60 = 166 lbs
If you are female:
100 lbs for 5 feet tall. Add 5 lbs for every inch over 5 feet.
Ex: Sarah is 5'5" tall.
Therefore her IBW is:
100 + (5 lbs x 5")
100 + 25 = 125 lbs

* Note: If you are less than 5 feet tall, subtract 2.5 lbs per inch less than 5 feet.

** The IBW formula is not for everyone. It is for average adults between the ages of 20 and 65 who want to determine their desirable body weight range. The IBW is not for pregnant or lactating women or for children. If you are concerned about your weight, please consult your health care provider.

If you are within 10% (about 10-15 pounds) of the IBW, you are in a healthy weight range. If you weigh more than 10% of your IBW, then you would be classified as overweight for your height. If you weigh more than 30% of your IBW, then you would be classified as obese. Being overweight or obese is a risk factor for many serious conditions, and weight reduction should be seriously considered for your overall health.

> **Figure out _your_ IBW! ****
>
> Your height = ___' ___"
>
> **If male:** 106 + (6 lbs x ___")
> Your IBW= ___#
>
> **If female:** 100 + (5 lbs x ___")
> Your IBW= ___#

Blood Pressure

The amount of sodium you take in through the foods you eat can directly affect your blood pressure. High blood pressure can lead to a number of health conditions, such as heart disease or kidney disease. High blood pressure can also influence brain health because of increased risk of strokes. Eating too much sodium-containing salt is one of the primary ways your blood pressure increases.

Sodium, most commonly found in table salt, is also found in extremely high amounts in food items that contain preservatives. Preservatives are used to keep foods looking and tasting fresh for any set period of time. Some of the items that contain high amounts of salt and/or preservatives include canned vegetables and soups, frozen microwaveable dinners, and smoked or processed meats, such as hot dogs, sausages, and lunch meats. Following a lower-sodium diet can be the key to helping relieve high blood pressure.

A Nutrition Facts Label Example

If you look at the Nutrition Facts label on the back of any food item, an ideal amount of sodium to look for would be 140 milligrams or less per serving. This would mean that the particular item would be considered "low sodium" and contains minimal amounts of salt and other sodium-containing preservatives.

Check out the Nutrition Facts label that follows. When reading a label, first check the serving size. The serving size differs from item to item so you will need to look at this for each food. On this particular Nutrition Facts label, the serving size is one cup. If you scan down to the sodium content, you will find that this particular item contains 660 milligrams of sodium per serving. So, for every cup of this item, this means you will take in nearly five times what is recommended per serving for sodium intake! If the item contained less than 140 milligrams per serving, then it would be considered a low sodium item and could be labeled as such.

Cholesterol

High blood cholesterol levels can lead to heart conditions and can also contribute to a decline in brain health. Cholesterol is a fatty substance closely related to the development of blocked blood vessels. Cholesterol comes from two sources: it is naturally produced by the body and it can come from the foods you eat.

Why does cholesterol influence brain health? Simply stated, the answer is high cholesterol causes an increase in buildup on the blood vessel walls. When this plaque builds up, there is less room for the blood to flow through your blood vessels, which leads to an increased risk of clots that can cause blockages to the heart (heart attacks) or to the brain (strokes). Remember, your brain demands 25% of the blood from each heartbeat.

Nutrition Facts	
Serving Size 1 cup (228g)	
Servings Per Container 2	
Amount Per Serving	
Calories 260	Calories from Fat 120
	% Daily Value*
Total Fat 13g	**20%**
Saturated Fat 5g	**25%**
Trans Fat 2g	
Cholesterol 30mg	**10%**
Sodium 660mg	**28%**
Total Carbohydrate 31g	**10%**
Dietary Fiber 0g	**0%**
Sugars 5g	
Protein 5g	
Vitamin A 4% •	Vitamin C 2%
Calcium 15% •	Iron 4%

*Percent Daily Values are based on a 2,000 calorie diet. Your Daily Values may be higher or lower depending on your calorie needs:

	Calories:	2,000	2,500
Total Fat	Less than	65g	80g
Sat Fat	Less than	20g	25g
Cholesterol	Less than	300mg	300mg
Sodium	Less than	2,400mg	2,400mg
Total Carbohydrate		300g	375g
Dietary Fiber		25g	30g

Calories per gram:
Fat 9 • Carbohydrate 4 • Protein 4

Lowering the amount of cholesterol you eat can aid in the prevention of these chronic illnesses. Cholesterol is only found in foods from animal products. Some examples are butter, milk, cheese, meats, and lard.

Take another look at the Nutrition Facts label. If you wanted to purchase foods low in cholesterol, you would simply look for items with 20 milligrams or less per serving. For this particular food item, each cup contains 30 milligrams of cholesterol, which does *not* make it a "low cholesterol" food.

Physical Activity

Physical activity is another key component of increased brain function because of the increased blood flow to the brain areas and the chemical response of the body during and after exercise. Exercise helps to release the "feel-good" hormone *endorphin*. This hormone has a positive effect on the body by helping to lower stress levels and improve self-esteem.

"Food for Thought"

After reading this section on nutrition and brain health, you might have noted an overall theme: eating more fruits and vegetables relates to brain health! The Dietary Guidelines currently recommend that most Americans should aim for "5-a-day" — about 2 cups of fruit and 2½ cups of vegetables per day for overall health. Seems simple enough, right? But, are *you* going to make the nutritional changes to improve your own brain health? Lifestyle changes are important and extremely *necessary* to enhance brain health and prevent chronic illnesses. Start now and reap the benefits of improved overall brain function.

References for Nutrition

1. Corrada MM, Kawas CH, Hallfrisch J, et al. Reduced risk of Alzheimer's disease with high folate intake: The Baltimore longitudinal study of aging. *Alzheimer's & Dementia* 2005;1:11-18.
2. Howitz KT, Bitterman KJ, Bohen HY, et al. Small molecule activators of sirtuins extend Saccharomyces cerevisiae lifespan. *Nature* 2003; 425:191-196.
3. Buhler DR, Miranda C. Antioxidant activities of flavonoids. Department of Environmental and Molecular Toxicology: The Linus Pauling Institute at Oregon State University. 2000 Nov. Available at: http://lpi.oregonstate.edu/f-w00/flavenoid.html. Accessed 2006 Aug 2.
4. Engler MB, Engler MM, Chen CY, et al. Flavonoid-rich dark chocolate improves endothelial function and increases plasma epicatechin concentration in healthy adults. *J Amer College of Nutr* 2004;23(3):197-204.
5. Grassi D, Necozione S, Lippi C, et al. Cocoa reduces blood pressure and insulin resistance and improves endothelium-dependent vasodilation in hypertensives. *J Hypertension*: Amer Heart Assoc 2005;46:398.
6. Wu X, Beecher GR, Holden JM, et al. Lipophilic and hydrophilic antioxidant capacities of common foods in the United States. *J Agric Food Chem* 2004;52(12):4026-4037.
7. Whitmer RA, Yaffe K. Obesity and dementia: lifecourse evidence and mechanisms. *Aging Health* 2006;2:571-578.

Additional Resources for Nutrition

American Dietetic Association www.eatright.org

5-a-day organization www.5aday.org

United States Department of Agriculture www.usda.gov

American Heart Association www.americanheart.org

Dietary Guidelines www.health.gov/dietaryguidelines

ADA "Evidence Supports Good Nutrition For Active Healthy Aging"

www.eatright.org/cps/rde/xchg/ada/hs.xsl/career_1693_ENU_HTML.htm

Three-Day Sample Menu Including Brain-Healthy Foods

Day 1	Day 2	Day 3
Breakfast 1 cup oatmeal cooked with soy milk and sprinkled with cinnamon 1 orange 4 oz unsweetened carrot juice **Snack** 1 plum **Lunch** 3 oz grilled chicken breast with topping: ½ cup onion slices and ½ cup green bell peppers, cooked down in 1 tsp canola oil 1 small baked sweet potato with skin 1 cup green beans cooked with 1 Tbsp lemon juice and 1 Tbsp almond slices Unsweetened beverage **Snack** 4 oz low-fat cottage cheese 1 peach **Dinner** Southwestern Shrimp Wrap: 3 oz grilled shrimp, ½ cup black beans cooked with 1 Tbsp lime juice & ¼ tsp turmeric (add cayenne pepper if desired) 1 cup shredded romaine lettuce, 1 whole grain flour tortilla wrap 1 cup sliced tomatoes 1 cup skim milk	**Breakfast** Sunshine Smoothie 1 cup frozen (or fresh) mixed berries, 4 oz unsweetened orange juice, ½ banana, 6-oz container light/non-fat yogurt and 2 tsp wheat germ blended with ice **Snack** ¼ cup pistachios **Lunch** 3 oz albacore tuna (fresh or canned) Cooked as desired 2 slices whole grain bread 2 cups raw spinach salad topped with 2 Tbsp dried cranberries and 1 Tbsp olive oil vinaigrette: 1 tsp olive oil 2 tsp vinegar black pepper to taste Unsweetened beverage **Snack** 1 cup cubed cantaloupe 1 6-oz container light/non-fat yogurt **Dinner** 3 oz grilled beef sirloin 2/3 cup steamed brown rice ½ cup cooked collard greens ½ cup cooked sliced beets 1 cup skim milk	**Breakfast** 1 cup toasted oats cereal 6 oz soy milk 4 oz unsweetened grapefruit juice **Snack** 2 kiwi fruits **Lunch** 3 oz roasted turkey breast 1 cup black-eyed peas ½ cup steamed broccoli 1 cup raw cabbage salad: shredded cabbage mixed with ½ cup raw shredded carrots and 1 Tbsp olive oil vinaigrette (see recipe Day 2) Unsweetened beverage **Snack** 2 Tbsp homemade hummus cooked chickpeas mashed and mixed with 1 fresh mashed garlic clove and 1 tsp olive oil then sprinkled with sesame seeds ½ whole grain pita bread pocket **Dinner** 3 oz grilled salmon 1 small potato with skin mashed or baked 6 asparagus stalks broiled until tender 1 cup skim milk

What Brain Research Teaches about RRR

About Lauren M. Dorgant

Lauren M. Dorgant is a Licensed and Registered Dietitian residing in Lake Charles, Louisiana. She obtained her Bachelor of Science degree in Dietetics from Louisiana State University and completed her Dietetic Internship at McNeese State University. Lauren is currently employed by a regional medical center in Lake Charles, where she coordinates patient nutrition education through outpatient clinics, including diabetes, congestive heart failure, hepatitis, oncology, as well as others. In addition, she teaches weight management classes, diabetes, and cardiac nutrition classes.

Ms. Dorgant serves on the medical center's Diabetes Advisory Committee and is a member of the statewide Disease Management Initiative for Diabetes. A member of the American Association of Diabetes Educators and the American Dietetic Association, she also currently serves as the president-elect of the Southwest chapter of the Louisiana Dietetic Association. In October 2006, she obtained her Certificate of Training in Adult Weight Management.

Being a Louisiana native, Ms. Dorgant realizes that food is an immeasurable component of our culture and that old habits are hard to break, especially changing the traditional dietary habits of southern families. With novel emphasis on diet as it relates to the important matter of brain health and functionality, Ms. Dorgant works to promote healthier lifestyles through improved food choices.

Appendix Three

Resources

Publications

Nussbaum, P.D. (October, 1999). Lifelong Learning and Wellness: One Component to the Enlightened Gerosphere. *Responses (11-13)*.

Nussbaum, P. D. (2001). *Alzheimer's Disease Virtual Preceptorship - CD*. Health Answers.

Nussbaum, P.D. (2002). Learning: Towards health and the human condition. *Educational Technology, 42*, 35-39.

Nussbaum, P.D. (2006). The brain-health benefits of lifelong learning. *The Older Learner, 13*, 4-7.

Nussbaum, P.D. (2007). Brain health and activity. *The Career Planning and Adult Development Journal, 22,* 24-35.

Presentations

Nussbaum, P.D. (March, 1999). *Putting the action into lifelong learning: From research to practice.* Presented at the 45th Annual Meeting of the American Society on Aging, Orlando, Florida.

Nussbaum, P.D. (June, 1999). *The Power of Learning across the Lifespan: Using the Brain for Wellness.* Presented at the Summer Series on Aging sponsored by the American Society on Aging's National Learning Center, Pittsburgh, Pennsylvania.

Nussbaum, P.D. (February, 1999). *Lifelong Learning and Brain Wellness.* Paper presented as part of a workshop entitled In Search of the Meaning of Learning at the Annual Convention of the Association for Educational Communications and Technology, Long Beach, California.

Nussbaum, P.D. (March, 2000). *Effects of Aging on Memory and Learning: Implications for Policy and Programming.* Presented at the 46th Annual Meeting of the American Society on Aging, San Diego, California.

Nussbaum, P.D. (October, 2000). *Lifelong Learning: Implications for Health and the Human Condition.* Paper presented at the

Presidential Session for the International Conference of the Association for Educational Communications and Technology, Denver, Colorado.

Nussbaum, P.D. (March, 2001). *Practical Implications of New Brain Research*. Paper presented at the 47th Annual Meeting of the American Society on Aging, New Orleans, Louisiana.

Nussbaum, P.D. (April, 2002). Invited Keynote Address: Mind Alert Lecture, *The Wonderful and not-so Mysterious World of the Human Brain*, 48th Annual Meeting of the American Society on Aging, Denver, Colorado.

Nussbaum, P.D. (March, 2003). *The Human Brain and How it Learns*. Presented at the Annual Conference of the American Society on Aging and the National Council on Aging, Chicago, Ilinois.

Nussbaum, P.D. (October, 2003). *Brain Health and Wellness*. Third Annual Harold B. Deets Lecture, MIT, Boston, Massachusetts.

Nussbaum, P.D. (July, 2004). *Brain Health Across the Lifespan*. Keynote address for the National Alzheimer's Association Annual Meeting, Philadelphia, Pennsylvania.

Nussbaum, P.D. (October, 2004). *Learning and Brain Health*. Keynote Lecture to Annual Conference of Osher Learning Institute, Portland, Maine.

Nussbaum, P.D. (March, 2005). *Preventative Aspects of Alzheimer's Disease*. Presented to the Aging Center and Geriatrics Division, Miami of Florida University, Miami, Florida.

Nussbaum, P.D. (September, 2005). *Brain Health and Education*. Invited presentation to the Osher Institute on Learning, University of Nebraska, Lincoln, Nebraska.

www.paulnussbaum.com